THE DESIGN OF
COMPLEX
INFORMATION
SYSTEMS

GRAYCE M. BOOTH

Honeywell Information Systems Inc.

THE DESIGN OF
COMPLEX
INFORMATION
SYSTEMS

Common Sense Methods
for Success

McGRAW-HILL BOOK COMPANY

*New York St. Louis San Francisco Auckland Bogotá Hamburg
Johannesburg London Madrid Mexico Montreal New Delhi Panama
Paris São Paulo Singapore Sydney Tokyo Toronto*

Library of Congress Cataloging in Publication Data

Booth, Grayce M., date.
The design of complex information systems.

Bibliography: p.
Includes index.
1. System design.   I. Title.
QA76.9.S88B658   1983   001.64   82-13080
ISBN 0-07-006506-3

1 2 3 4 5 6 7 8 9 0   K P K P   8 9 8 7 6 5 4 3 2

ISBN 0-07-006506-3

The editors for this book were Stephen G. Guty and Diane
Krumrey, the designer was Elliot Epstein, and the production supervisor was Paul Malchow. It was set in Baskerville
by Bi-Comp, Incorporated.

Printed and bound by The Kingsport Press.

CONTENTS

PREFACE

The data-processing environment is changing rapidly, and while this adds to the excitement associated with this dynamic field, it also increases the challenge to remain up to date. Distributed processing, word processing and other office-automation facilities, personal computers, and techniques such as teleconferencing and viewdata all represent possible ways to accomplish the information-processing tasks of an organization.

Unfortunately, we too often become so involved with these technical complexities that we lose sight of what the real goal is: to assist the organization in carrying out its tasks more efficiently and effectively. To do so, it is necessary to have an in-depth understanding of what those tasks are, how the organization is structured to accomplish them, and what kind of background the users have. In short, it is essential to be able to accurately define the requirements for an information-processing system, and to design that system to fit comfortably within the organizational environment.

This book was written because of my increasing conviction that many organizations have data-processing problems because they concentrate too much on technical details and not enough on system analysis and design. There is often a great deal of pressure to implement a new application quickly, and the most visible activity during implementation is programming. The "front-end" functions of requirements definition, system analysis, and system design are therefore often done hurriedly—or sometimes not at all—so that programming can begin. The resulting system may be completed on time (although often it is not), but it probably will not meet the real needs of the organization. In contrast, in my experience those organizations which have spent a great deal of time defining *what* is needed, and in building a prototype of the proposed system so that the users can experiment with it before the design is finalized, have been far more successful in meeting their goals.

The focus of this book, therefore, is on the definition of requirements, and the process of system analysis and design to meet those requirements. This is not a theoretical work, but rather one resulting from practical experience and, as the subtitle indicates, the application of common sense to complex problems. The reader will find workable procedures and guidelines which can be put to use immediately, and easily, no matter what type of organization or application is involved, and no matter what kind of computer equipment will be used. I have emphasized the need for communication among the users, the functional management of the organization, and the data-processing staff, because effective communication is the key to good requirements definition and system design. If the data-processing professionals and the users insist on speaking different languages—"DP" jargon in the one case, and English or possibly jargon from another discipline (law, medicine, etc.) in the other—it will be impossible to design and implement satisfactory systems.

The 1980s will present many challenges, as we collectively learn how to apply new technologies to both old and new problems. I believe, however, that the greatest challenge is for data-processing professionals and the users of data processing to learn how to communicate more effectively than has often been the case. I hope that this book will assist the reader in meeting that challenge.

Grayce M. Booth

THE DESIGN OF
COMPLEX
INFORMATION
SYSTEMS

INTRODUCTION

1

TRENDS IN
SYSTEM ANALYSIS
AND DESIGN

The process of system analysis and design is fundamental to the creation of the computer-based information systems which are increasingly important to the operation of many organizations. Although system design is often discussed, this term—like many others used in the computer industry—has many different meanings. It is important, therefore, to define system analysis and design within the context of this volume.

This book deals with computer-aided application systems. (The term "computer-aided" is used in preference to "computerized," to indicate that applications are rarely completely automated, but instead involve interaction between people and computer-based logic.) The implementation of the operating systems, compilers, and so on which support these applications is outside the scope of this work. For convenience, the term "system design" will generally be used, as this is the common way of referring to the design of any type of computer-based process.

The implementation of a computer-aided application begins with a definition of the requirements, which can best be expressed as a set of functions to be provided. Since these functions must be used by people, a definition of the interfaces between the users and the computerized functions is also required. Most often these interfaces take the form of terminal-usage procedures, but they may also include "interfaces" which are reports produced by the system for its users.

The major emphasis in system design is usually on selecting the methods by which the needed functions will be provided. Specific hardware devices—such as information processors, peripheral devices, terminals, and so on—must be selected if sufficient equipment of the appropriate types is not already installed. The necessary software functions must be defined, and either obtained from an outside source or designed in-house. This level of software design involves separating the necessary

functions into specific programs, and possibly into modules within programs.

These steps, which begin with the definition of the functional requirements and end with the top-level software design, form the system analysis and design procedures which are covered in this book. Of course these processes are typically followed by more detailed system design and by software implementation, either for the total system or for phased subsets. When the coding is complete, each set of software must be tested for correct operation. If the system includes multiple sets of application software, as is typically the case, all the sets must be tested together to ensure that they operate correctly in concert. System evaluation also includes testing of the user interfaces, to make sure that they operate as expected and are in fact well suited to the needs of the users. Upon completion of the test phase, the necessary hardware and software are installed for production use. This may be a one-time process, or may involve a gradual phase-in of different user groups, different locations, and so on.

This volume is concerned with the system analysis and design phases of the overall process because these are so important to the success of complex systems. Too much emphasis is often placed on the detailed system design and implementation phases. These are, of course, technically complex, but skill in these phases is of no use if the system requirements are not well understood and/or the overall design of the system is poorly chosen or poorly structured. A technically elegant implementation of functions which no one really needs, or which are difficult and cumbersome to use, is not a successful system.

It is interesting to note that the analysis and design of computer-based information systems are very similar to the design of manual information-processing systems. In each case, the functions to be performed must be defined and methods for performing them must be specified. Interfaces between the various elements of the information system must also be defined. In a manual system these interfaces are between people; in a computerized system interfaces exist between people and machines, between people, and often also between machines. In any complex system, whether based on the use of computers or on the use of people as "processing elements," the most difficult task is deciding what processing to do, when to do it, and how to do it. It is unfortunate that the introduction of the computer led to the expectation that it would eliminate many of these problems through its superior calculation and logic capabilities as well as its impressive speed. (Although computers have been in widespread use for close to 30 years, this expectation persists.)

In practice, computer-based information systems have proved to be even more difficult to analyze and design than people-based information systems. This is largely because computers must be told unambiguously exactly what to do—their programs must deal with each possible situation which can arise, and must define exactly what is to be done in each case. People, in contrast, can work with ambiguous procedures—within reasonable limits—because they can analyze an unusual situation and determine what to do even if their operating procedures do not define the necessary actions. Typically, therefore, operating procedures for manual systems are less than complete—indeed in many organizations written procedures are nonexistent. On-the-job training by fellow workers or supervisors, followed by experience, is expected to combine with the worker's inherent intelligence to produce the desired result—a person who knows what to do in all, or at least the majority, of the situations encountered in day-to-day operations.

Many people, including many computer-application designers, have little or no experience in defining unambiguous and complete operating procedures. This made relatively little difference when most computer-based systems were used for batch processing, since early batch-mode applications were concentrated in the financial areas where methods *were* relatively well-defined because of the financial implications of ambiguous procedures. It was therefore not too difficult to implement a batch-processing system to handle payroll, accounts payable, accounts receivable, and similar applications.

Data-processing people in general sustained a considerable shock when they attempted to transfer this experience into the on-line world of applications such as order entry. The procedures for taking and processing orders in most establishments were far less well defined than the financial procedures. A company might have an order-entry form, but often not all orders were recorded on the form, and each salesperson or order clerk might fill out the form somewhat differently. As long as the forms and the order processing were handled by people, this level of ambiguity was acceptable, and only a tolerable number of errors (in most cases) occurred.

Attempts to computerize procedures of this type often, in effect, involved rethinking and redesign of the entire business process—with all the attendant upheavals, personnel discontent, and perhaps organizational restructuring of the people involved. On-line systems which have been termed "failures" have as often (if not more often) suffered from problems of this type, inadequately handled, as from technical problems.

NEW TYPES OF APPLICATIONS

Many organizations today place great emphasis on skills in system analysis and design. This emphasis is the result of a shift in the types of applications which are being selected for computerization. The majority of the applications which operate well in batch mode have been implemented by now, and most organizations are moving into the implementation of increasingly complex new applications. While those applications are of many types, in many different organizations engaged in business, government, education, health care, etc., they generally share the following characteristics.

ON-LINE INTERACTIVE PROCESSING

On-line, interactive processing is typical. Either these are new applications, which have never before been assisted by computerized logic, or they are major revisions of existing applications designed to move them from the batch environment to the interactive environment.

END-USER ORIENTATION

End-user orientation is extremely important because these new applications are intended to serve people who have little or no data-processing experience. Although the use of personal computers is growing enormously, the vast majority of the people in the United States (and in other countries) today have had no experience with any type of computer. This lack of computer knowledge will not disappear, even as new generations graduate from school with some level of computer experience, because many people simply have no interest in computers. (Some professionals in data processing believe that everyone would like to program and use a computer if they had the opportunity, but this is a biased view. People who are intensely interested in computers are not representative of the population as a whole.)

End-user orientation therefore requires that the computerized functions act as a tool for the user; they must be available whenever needed, easy to understand and to use ("friendly"), and responsive to the user's requirements.

INTEGRATION OF APPLICATIONS AND FUNCTIONS

Integration of applications and functions is increasing. Most organizations today operate compartmentalized data processing, office systems, data communications, voice communications, manufacturing systems, etc. Current advances in both information processing and communications technology make it increasingly attractive to integrate more and more of these functions, both technically and organizationally. Large businesses, in particular, are forming new organizations called information resource management, or IRM,* which are responsible for all the automated and manual procedures and communications techniques within the company. This trend, which is still in its early stages, represents a desire to capitalize on technological advances to improve productivity and management control.

An area of particular interest, and one which will be dealt with in some depth in this volume, is the merging of "classical" data processing with office-automation functions. ("Office automation" is a poorly chosen term, because automation—in the same sense as factory automation—is clearly impractical at present and may never become practical.) Office-support devices, which began as free-standing productivity aids for secretarial and clerical workers, are rapidly evolving into distributed systems linked by electronic-mail facilities. It has been predicted frequently that the really significant productivity improvements in an office will come when computerized aids are provided to managers and professionals. These people have extremely varied requirements, ranging from voice mail and teleconferencing to the ability to easily access computer databases to run "what if?" simulations of future conditions. The trend is clearly toward distributed systems which include both data-processing and office-support capabilities; in fact, the distinction between the two is quite arbitrary and will no doubt disappear from the industry's vocabulary in the future.

MAIN-LINE INFORMATION SYSTEMS

Computer-based information systems which support main-line functions are becoming typical, although not many organizations are really "run by computers" as yet. Airlines are perhaps the prototypes of how many organizations will operate in the future. The main-line functions of the airlines, which include selling tickets, making seat reservations, and scheduling equipment and personnel, are heavily computer-aided, and in fact depend entirely upon the operation of computer systems. Increasingly automated factories are moving manufacturers toward the same dependence upon computer-based systems. Many other types of organizations are reaching the point at which they are dependent upon the continued, and correct, operation of their computers, not just for after-the-fact reporting and financial processing, but for their moment-to-moment functioning.

This trend reflects the massive changes which have occurred in society in the United States as well as in other developed countries. Originally,

* The term IRM is used frequently in this volume to refer to the organization previously called the data processing (DP) department.

the United States was a farming-based society, in which a majority of the workers were involved in farming. As the productivity of farmers increased, the United States evolved into a manufacturing society. The largest single class of worker was no longer the farmer, but rather the factory worker. Quite recently, the United States has evolved into what has been termed an "information society," in which the largest single class of worker is the clerk. In fact, a great many organizations no longer produce a product—in the sense that a manufacturing organization does—but instead they produce and/or process information as a service. Examples of organizations based on information processing include banks, stock brokerages, the insurance industry, and many governmental organizations.

This trend intensifies the need for computer-based information systems to help improve productivity and lower costs. Information processing is highly labor-intensive when done manually, and productivity can best be improved by investing in tools which reduce the manual content of tasks. This does not necessarily mean that fewer people will be involved, but rather that more tasks can be done, or that tasks can be done more correctly and/or more rapidly, than before.

One of the most important results of the movement of computer-based systems into the mainstream of many organizations is that these systems must be extremely responsive to changes in an organization's environment. One well-known data-processing syndrome—in which a change to a computerized application is requested, only to receive the response that it will take 6 months (or even longer) to implement the change—is totally unacceptable in these cases. If a business uses an on-line order-entry system and the management decides to implement a new method of selling, the order-entry system must support that decision. If a business is under competitive pressure and must restructure its product line and pricing to combat this pressure, its computer-based system must respond rapidly to this urgent need.

This environment makes it essential that information systems be flexible, adaptable, and amenable to rapid change. Unfortunately, many computer-based systems today tend to be relatively inflexible and resistant to change. One of the major challenges for designers of information systems today is to devise ways to provide systems which are truly responsive to real-world changes. This theme will be repeated frequently throughout this volume, since it is one of the most important aspects of information systems of the 1980s and beyond.

BASIC GROUND RULES

This book is built around the use of several ground rules for the successful design of computer-based systems:

1. Involve the people who will use the system, as well as the management to whom those people report, in the process of system analysis and design. This does not mean that the users must become data-processing experts; however, they are typically the only people who fully understand which functions are required to carry out their jobs. Computer-based tools intended to aid those people must be designed with an in-depth understanding of what they do and how they do it. The process of enlisting the users will not only help to ensure that the design correctly reflects the work flow and the functions but will gain the sympathy of the users and their management. This is extremely important, as hostile users or unsupportive management can sabotage a system as effectively, if not more effectively, as technical data-processing problems can. If the system is a large, complex one, or one which will affect the main-line operations of the organization, it may be necessary to enlist the support of the users' management at a very high organizational level.

2. Define the goals and objectives of the system as completely as possible before beginning its design. An elegantly designed computer system which does not do what the organization needs done is a failure—no matter how technically sophisticated the implementation may be.

3. Obtain agreement, in writing, from the users and their management that the defined goals and objectives are appropriate. In addition, define specifically and formally how the attainment of those objectives will be measured. For example, if it is agreed that 2-second response time is needed for 90 percent of all transactions, define and agree upon a test period during which response time will be measured and state exactly how it will be measured. Again, explicit definitions of objectives and their measurement will ensure that the system really meets the needs of the users and the organization.

4. Don't view the system objectives, even when agreed to in writing by users and user management, as cast in concrete. The ability to match real-world changes is essential in today's information systems. It is important in any design to identify functions and requirements which are basic and will not change; for example, recording orders, billing customers, and paying employees are functions which will always be required. The specific methods used may change—even change dramatically—and particular items, such as prices, discount percentages, rates of pay, etc., are always subject to change at short notice. Every system must be designed to accept change readily.

5. Realize that prospective system users, unless they are very familiar with computer-based systems from past experience, are often unable to visualize what such a system can do. They may therefore be unable

to participate meaningfully in the process of setting objectives and analyzing system requirements. If this is the case, consider building prototypes of the system, or of the user interfaces to the system, so that the users will be able to see what the system *could* do. Often this is the best way—sometimes the only way—to obtain suggestions from the users as to what the system ought to do to be most helpful.

6. Organize the implementation and installation of the system in manageable phases with formally defined checkpoints, at each of which specific tasks must be accomplished. Too many large projects are undertaken with the idea that the system must be implemented as a whole, even if this will take several years.* This approach has two potentially severe disadvantages:

- The system may reach the "95 percent complete" mark, but never advance beyond that state because of the difficulty of determining exactly how much remains to be done. The larger and more complex the system, the more likely it is that this problem will be encountered.

- The requirements of the system users may change substantially during such a long implementation process, so that even if the completed system exactly matches the original specifications it may no longer perform any useful functions.

Both these problems can be avoided by defining reasonably short and manageable implementation phases, each of which is easily measurable and produces something usable. In general, phases between checkpoints should not exceed 6 months, and for project-management purposes it is preferable that the time periods be as short as possible.

It may be noted that these ground rules have very little to do with technology or with methodologies for *computer-system* design. Instead, they are mainly concerned with the relationship between the users of the system (and their management) and the designers of the information system—and especially with the communication between these groups. Although the design of a complex computer-based system is a formidable technical problem, failures in systems of this type are most often caused by the difficulty of deciding exactly what the system ought to do—and this in turn is caused by the difficulty of communication be-

---

* One organization began the implementation of a computer-based system with the expectation that there would be no usable result for 8 years! No situation remains stable long enough for this to be a rational plan.

tween the system users and system designers. Improving communication is therefore a central theme in this book.

SYSTEM ANALYSIS, DESIGN, AND IMPLEMENTATION

To define the process of designing and implementing a computer-aided application system in more detail than the earlier brief summary, the following eleven steps are required:

1. Defining potential applications—usually in response to some current or potential future limitation of manual procedures, and/or because of a general belief that the computerization of specific functions would improve cost, productivity, or effectiveness.

2. Evaluating the return on investment (ROI) of various potential applications and selecting one or more for further analysis. For the purposes of this discussion, it will be assumed that only one application at a time is selected for further study.

3. Defining the objectives for implementation of the selected application in a computer-aided form. This involves determining how the potential users of the new system can be helped, which defines some of the objectives. Additional objectives may be stated by management; for example, increased competitive strength through faster analysis of market trends might be a management objective.

4. Collecting data which describe the characteristics of the application. This is the first part of the system-analysis process, and it involves a study of who will use the new application, how it will be used, what the requirements are for terminal-response time, and so on.

5. Analyzing the collected data. This is also included in system analysis, but forms the transition to system design. Through analysis of the data which describe the prospective application, many operational and organizational patterns can be detected. These in turn may define, or at least heavily influence, the design of the computer-aided system.

6. Carrying out the high-level or strategic system design, in which decisions are made regarding which functions to implement, which to include in hardware and which in software, and making trial designs of the user interfaces to the system.

7. Evaluating the system design to this point, to determine that it not only is technically sound but will meet the objectives defined. If this

evaluation produces a negative result, either an iteration through earlier steps in this sequence is required or a decision must be made that this is not a good candidate application for implementation—at least under the present circumstances. If the results of the evaluation are favorable, then a management decision can be made to continue into the detailed design and implementation steps.

8. Carrying out the detailed system design, which includes selecting the hardware and software elements to be used. It also includes the design of the parts of the software which are to be implemented in-house. This step may also include an analysis of the implementation phasing to determine the sequence in which parts of the system—especially if it is large and complex—will be implemented. (This phasing decision is included in step 7 in many cases.)

9. Coding the necessary software or that subset which is selected for the initial phase of the implementation.

10. Performing test and qualification of the software (really for the combination of hardware and software), first as individual software elements (modules, programs), then with elements grouped into components, and finally of the system as a whole. Successful completion of this step indicates that the system can be expected to operate correctly and that it has been accepted by the users and their management as ready for installation.

11. Cutover of the system to production mode, usually in phases—by location, by function, or segmented in some other way—to make installation easier. This cut-over phase ought to be followed by a postinstallation audit or survey to determine if the system is performing as expected, meeting the defined objectives, and providing the benefits which had been forecast during the early steps of this process.

Reducing this system analysis, design, implementation, and installation process to eleven steps may make it seem relatively simple, but of course in a complex computer-based application each of the steps may be quite time-consuming and difficult. This is one of the major reasons for partitioning complex systems into manageable parts and treating each as a relatively independent entity during design and implementation.

This volume provides an overview of steps 1 through 3 of the above process. Since the exact details of these steps are heavily dependent upon the type of organization and the types of applications being considered, it is difficult to define a detailed procedure which will apply in all cases.

However, certain actions must be taken in steps 1, 2, and 3; those actions are described in Section 2 of this volume. Steps 4 through 7 form the heart of the strategic system analysis and design process. They are the focus of this volume, and each is covered in a major section (Sections 3 through 6). Steps 8 through 11 form the detailed design, implementation, and installation phases and are not covered (except in references to certain aspects of those steps) in this volume. As was noted earlier, much of the emphasis in computer science literature has until recently been on these detailed steps and the various techniques to be used. It is therefore unnecessary to treat them here.

Because the problems of system analysis and design are so complex, no single volume can realistically promise to solve all those problems by presenting a foolproof method for the design of complex systems. This book, which is based upon a considerable amount of practical experience (both personal and vicarious), is intended to provide basic approaches and methods. Readers may wish to modify and/or extend these on the basis of their experience, particular areas of expertise, and areas of effort.

In addition, no single volume can adequately cover all facets of the design of all possible types of information systems. As the title indicates, this book is about complex systems, with special emphasis on distributed systems. The methods described here are not applicable to simple, single-function systems or to most single-application systems—those can be designed and implemented using simpler methods. Complex systems, in contrast, amply repay the extra time, effort, and organized methods used during the analysis and design process.

PREPARATORY WORK

2

SELECTING
CANDIDATE
APPLICATIONS

Every organization has some way of defining which new applications will be implemented in computer-aided form and which existing applications will be expanded or modified. This discussion excludes application changes which are typically called *maintenance*, i.e., which involve relatively minor modifications. For example, changing payroll programs because the rate of social security deductions changes at the beginning of a year is the sort of (usually) minor change which goes on continually. This chapter covers major extensions or modifications, such as moving a batch-mode application to on-line mode or adding a number of new functions to an application.

It has been estimated that the typical IRM department today has a backlog of major modifications and potential new applications which totals 2 or more years of effort for the staff. Because the typical staff spends more than half their time on routine maintenance activities, this backlog represents a difficult management problem. Program maintenance often cannot be avoided or postponed, and yet the inability to implement new applications is a tremendous handicap to many organizations. Since the problem of program maintenance is unlikely to disappear, it is essential that the effort available for the development of new applications be directed toward the most important tasks.

Unfortunately, many organizations find it hard to decide which *are* the most important new applications. Effort may therefore be expended on work which may have the highest level of visibility but is not of the highest priority. To avoid this situation, it is necessary to define how applications are to be objectively evaluated and selected for implementation.

Each organization ought to maintain a pool of applications which are candidates for implementation. Entries can be made in this candidate list in any or all of the following ways.

- The users (or user management) of an existing application may describe enhancements which would improve the system's usefulness.

- Management (at any level) may define target areas of implementation, with specific goals such as improved productivity, lower cost, greater efficiency, or quicker service to customers or clients.

- System analysts may define major enhancements which would improve an application's responsiveness, performance, availability, or cost-effectiveness.

- External sources—such as the national or local government or an industry association—may define new laws, new standards, or new methods of operation which dictate changes in existing procedures. This may cause corresponding changes in computer-based applications, and/or may effectively force the implementation of new applications.

The list of potential new applications and enhancements may therefore consist of two types of entries:

1. Tasks which *must* be done because of legal, contractual, or similar situations.
2. Tasks which are, at least potentially, desirable to do.

The tasks in category 1 must be put immediately on the schedule of work to be done. Those applications or changes can bypass the study described in the remainder of this chapter, and the definition of objectives described in Chapter 3 can begin at once. For all tasks in category 2, however, it is necessary to make an initial determination of which tasks are most worthwhile. The rest of this chapter covers the steps necessary to make a first-cut evaluation of which new or enhanced applications are potentially most valuable and therefore worth more detailed study.

EVALUATING RETURN ON INVESTMENT

One method of deciding which applications to work on next is to do a preliminary evaluation of the expected ROI. (The term ROI is used here in a general sense, without the exact connotations it carries in connection with a company's balance sheet.) To compute an application's ROI, the following information must be obtained:

• The expected value of the results which the application will provide.

• The expected cost of the implementation and subsequent operation of the application.

   If this initial phase is to be kept within reasonable limits, these questions must be answered by performing a brief, top-level analysis. Although this approach is subject to error, as long as the same procedure is used for the preliminary evaluation of each candidate application the results—and therefore the comparisons based on those results—ought to be consistent.
   The expected value of an application can often be expressed in monetary terms, although this is not always practical. Some methods of defining the expected value are as follows.

• An application which will improve the productivity of some part of the work force will produce a value by allowing operation with fewer people in the work force, and/or the performance of more work with the same number of people. In practice, work force size is seldom reduced, except possibly by attrition (although work-force reductions are more common in the United States in the 1980s than they were in the sixties and seventies). However, productivity improvements often allow a given work force either to produce more work of the same kind or to perform new, additional types of work. If the cost per operation within the work force is known, or can be computed, then a value can be calculated which represents the potential improvement available with implementation of the new application.

• An application which will provide information not previously available will also be of value, but this value may be quite difficult to quantify. If, for example, having additional information concerning market trends would make it possible for a firm to be more competitive, it might be possible to assign a value to that information. In other cases, it may be extremely difficult—or effectively impossible—to do so. If the prospective users of this information cannot assign a monetary value to it, they must be strongly encouraged to at least decide how urgently the information is needed. If it is impossible to assign a monetary value or an "urgency value" to the results of a new or enhanced application, a low priority for implementation will result. This represents a potential problem, since the application might have a very significant positive impact on the organization which cannot be realized because of the difficulty of calculating a value associated with that impact.

• An application which will support the offering of new products or services to an organization's customers or clients can be assigned a value based on the expected profitability of the product or service. In the case of a nonprofit organization, it is again necessary to assign some monetary or urgency value to the support of new services (unless these are legally mandated), so that the application's priority can be defined.

There may be other ways to define the expected value of a new or enhanced application. In every case, it is extremely desirable to assign a monetary value if this is practical. However, if it is impractical or too time-consuming to do so, then an urgency value must be assigned. In this early stage of the system-analysis process there is a delicate balance between spending too much time on defining an application's potential value objectively and in detail and spending too little time and effort, thus possibly arriving at an assessment of value that is too subjective. If too much time is spent on each application when there is a large backlog to be studied, only a few applications will be reviewed. It is important, therefore, to allow enough time to study each candidate application at approximately the same level of detail.

At this point in the application-selection process it is also necessary to make an estimate of how much the implementation or enhancement of the application will cost. This can be accomplished by making a general analysis of the amount of implementation effort involved, often through analogy with other completed applications. The effort needed can then be used to compute the estimated cost of implementation. The degree of technical risk involved must also be estimated and taken into account in analyzing the cost. If, for example, the application is very similar to others which have been implemented by the same organization, the technical risk is likely to be minimal and the cost can probably be forecast quite accurately. However, if a new application requires the establishment of a database but there is no in-house expertise or experience with database management, a technical risk exists. High technical risk not only increases cost but also decreases the accuracy of the initial cost estimates. As with the definition of expected value, the evaluation of risk is often difficult and subjective. However, this evaluation must be made as objectively as possible—and then used to improve the accuracy of the implementation-cost estimate.

To compute the ROI of the application, both the one-time cost of implementation and the running cost of operation are needed. Most IRM organizations have established operational-cost rates which can be used as the basis for this calculation. Typically a life-cycle cost is computed to cover a period of 5 or more years (or less if the application is expected to have a short life cycle). Then the payback value of the application, as described earlier, can also be calculated for the same length of time. Of course the payback period does not begin until the application has gone into full production use. Start-up costs which precede this point must be taken into account in the calculation of ROI. So, in many cases, costs extend over 5 or 6 years, but the potential payback is available for only 3 or 4 years. Figure 2-1 shows a simple method for laying out estimated costs and payback value for ROI calculation.

If it is practical to assign a value to the application's results and also to calculate an expected life-cycle cost of implementation and operation, a straightforward computation of ROI is possible. Of course this computation can be made somewhat more accurate, and more complex, by using discounted-cash-flow methods, estimates of the future cost of money, etc. It is important to remember, however, that exact accuracy is less important than consistency among all calculations made at this time in the evaluation cycle. ROI computations at this point are used to define implementation priorities for the applications in the backlog—not to set budgets or realistic payback figures.

When this first ROI estimate has been made, some applications can immediately be given a priority within the candidate-application pool. Those applications which have a low ROI in comparison with the other candidate applications can be placed toward the end of the candidate list. Those which have a high ROI, either in comparison with other candi-

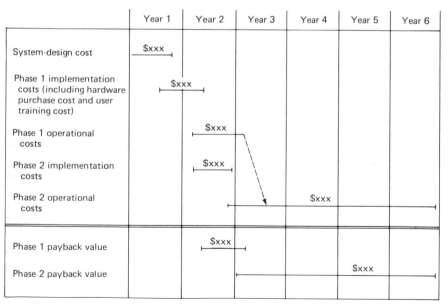

	Year 1	Year 2	Year 3	Year 4	Year 5	Year 6
System-design cost	$xxx					
Phase 1 implementation costs (including hardware purchase cost and user training cost)		$xxx				
Phase 1 operational costs			$xxx			
Phase 2 implementation costs			$xxx			
Phase 2 operational costs				$xxx		
Phase 1 payback value			$xxx			
Phase 2 payback value					$xxx	

Figure 2-1   Return-on-investment computation.

2. Does it serve one of the highest-priority organizational areas?

3. Will it fit within the organization's philosophy of centralized versus local processing functions?

Generally speaking, only candidate applications which produce positive answers to these questions are worth evaluating in more depth. In fact, if an application appears to offer an attractive ROI but does not fit well into the long-range plan, it may indicate that the long-range guidelines need some adjustment. This does not necessarily mean that there are errors in the long-range plan, but rather may reflect the fact that conditions change and that plans must be modified periodically because of those changes.

If a proposed application does fit into the long-range intentions of the organization and also shows a good ROI or has a high urgency, it is an attractive candidate for further study and potentially for implementation. The result of this first phase of application evaluation is to organize the application backlog into a general priority sequence. The resulting sequence will form two groups:

1. Applications which will support the highest-priority objectives and organizational or functional areas. Within this group, candidate applications will be arranged in sequence, with those representing the highest ROI or urgency at the top of the list.

2. Applications which are not in the highest-priority areas, similarly arranged by ROI and/or urgency.

Realistically, the sequencing of candidate applications within group 1 will be somewhat subjective. As discussed in the preceding section, it is impossible to compute ROI in some cases, and even when a value figure can be assigned, it is an estimate. In many organizations, there will be a large enough pool of candidate tasks in group 1 to keep the development staff occupied full time; in that case, no prioritization of the group 2 tasks will be needed. However, there ought to be a periodic reevaluation of the lower-priority candidates, as changing circumstances may cause an application's priority to increase and move it into group 1. Newly defined candidate applications must also be studied, their ROI or urgency determined, and a priority assigned. Some organizations form a team which meets regularly to evaluate new candidates and to periodically reevaluate the backlog for changing priorities. The use of a team, assigned specific responsibility for managing the application backlog, is a good idea—it ensures that adequate attention will be given to the backlog.

The typical backlog of unimplemented applications in both group 1 and group 2 presents a problem in many organizations. The lower-priority candidate applications may be of considerable value to specific users, even though they may be of less worth to the organization as a whole. The inability to implement those applications may not only hinder the effective operation of certain parts of the organization but may also create a significant public-relations problem between those user groups and the IRM staff. In many cases this situation is simply accepted as unavoidable. The users who do not receive service will generally take one or more of these four steps:

- Continually lobby for higher priority, to cause the implementation of their tasks.

- Give up and live with the situation, but unhappily.

- Obtain computing services from an outside source such as a service bureau.

- Buy their own small computers and implement applications locally.

Each of these actions presents a potential problem, and each organization ought to evaluate which potential problems are most manageable—and attempt to guide users toward those preferred actions. Ignoring the problems caused by the application backlog will not cause them to go away, but recognition of the difficulties, accompanied by enlightened actions to minimize the negative effects, can improve the organization's total information-systems strategy.

3

DEFINING
APPLICATION
OBJECTIVES

The evaluation of candidate applications, as described in Chapter 2, will very likely produce multiple candidates with high priority and attractive ROI. One or more of these can then be selected for more detailed analysis, beginning with the definition of objectives as described in this chapter. In a large organization, multiple applications may be analyzed simultaneously. For purposes of simplicity, however, the assumption is made here that only one candidate application is studied at a time.

The first step in system analysis is to define the objectives to be met when implementing the application. Like most of the other aspects of system analysis, the definition of objectives is an iterative process. All that can be done initially is to define the major goals, which can be termed strategic objectives. The definition of more detailed objectives may also begin at this time, but will typically be expanded and refined during later system-analysis phases.

DESIGN APPROACHES

There are two quite different ways to approach the design and implementation of computer-aided information systems, and the approach to be used ought to be decided upon when objectives are defined. The two contrasting methods are:

- The *grand-design* approach, in which a large, complex information system is designed and implemented as a whole.

- The *piece-at-a-time* approach, in which a general direction is established and then specific functions are designed and implemented as relatively independent building blocks.

THE GRAND-DESIGN APPROACH

The grand-design approach is the one most often favored by data-processing professionals. An integrated design can ensure that all functions and all parts of the system fit together well. In addition, in-depth analysis of the entire system prior to implementation ought to minimize the possibility of "surprises"—discovering new requirements after implementation has begun, or even after it is complete. Also, of course, major implementations present major technical challenges and are therefore attractive to many innovative system designers. However, this approach runs the risk of a very long design and implementation cycle, during which the requirements may change. In addition, the complexity of a large, integrated information system may extend the implementation and system-test time well beyond expectations. Meanwhile, the users who are waiting for the benefits of the new application(s) receive no results and may become disillusioned with the entire project.

Perhaps the greatest potential disadvantage of the grand-design approach is that it may cause traumatic organizational shifts. Suppose, for example, a manufacturing firm decides to ask its sales representatives to enter orders directly, via portable terminals. This may cause dissatisfaction among the sales staff, and it will mean that the order clerks who formerly handled order input and processing will have to be reassigned. Further, procedures that have been used for after-the-fact recording of orders by a dedicated staff will probably not be suitable for the new operation, which involves entering orders as they are taken. Procedures and organizational roles will therefore change simultaneously, with a high potential for serious difficulties.

This does not mean that the grand-design approach is never workable; there are circumstances in which it is quite appropriate. Usually these are implementations of well-understood functions, with no—or a minimal amount of—accompanying organizational change. This combination of well-defined functions and organizational stability minimizes the risk of failure. The grand-design approach may also be chosen if an organization's management believes that computerization, accompanied by changed methods (and perhaps by organizational changes), is essential. This is a valid management decision, but it ought to be made only with a full understanding of the risks involved.

THE PIECE-AT-A-TIME APPROACH

The piece-at-a-time approach, in contrast, is often disliked by data-processing personnel but appreciated by system users. A serious potential disadvantage of this approach is that individual system elements may be inappropriate to long-term goals and may not fit together well as the total system evolves. Individual elements may also be suboptimized, leading to inefficient operation when later combined with other elements. System users and their management generally like this approach because they receive some payback in system benefits relatively quickly. They are also less likely to face traumatic changes as each piece of the total system is implemented. If changes in operational methods and/or the structure of the organization are required, it may be possible to make these modifications gradually.

The piece-at-a-time approach is particularly advantageous—and the grand-design approach especially risky—when implementing computerized aids for poorly understood functions. An example application in this category is office support. How office workers, especially professionals and managers, accomplish their duties is poorly understood, and it is impossible to automate processes which are not understood; it is even difficult to support these processes with computer-based tools. If a major restructuring of office methods is undertaken as part of an office-automation project, the risk is high that the new methods will be no more effective than the old ones, but will be more expensive because of the investment in computer equipment.

The best way to tackle such projects is to set some general strategic goals, but to begin system implementation in a small-scale way and in close cooperation with the users (managers, secretaries, professionals, and clerks in the case of an office-based information system). As each system facility is implemented, obtain feedback from the users as to its good and bad features. If a rapport has developed between the users and the system designers, the users will quickly begin to provide guidance on how best to proceed toward achieving mutual goals.

Although the piece-at-a-time approach to design and implementation can cause a longer application-development cycle, the probability of creating a truly successful system may be higher with this method than with the grand-design approach. It is essential, however, that a well-understood set of strategic objectives be defined and kept clearly in focus during design and implementation. This will reduce, although not eliminate, the risk that the system "pieces" cannot be formed into a coherent whole.

TYPES OF OBJECTIVES

When defining objectives for a new or significantly enhanced application, it is important to clearly differentiate between the two types of objectives, strategic and implementation-oriented.

STRATEGIC OBJECTIVES

Strategic objectives define the desired end result for the organization. For example, the purpose of the application may be to lower costs (directly or indirectly), to improve efficiency, to improve the organization's competitive position, or to offer new services to customers or clients.

One or more objectives of this type ought to be defined for every new or significantly enhanced application. In general, this type of objective is defined during the ROI evaluation described in Chapter 2, since the strategic objectives define the value of the proposed application implementation. As Chapter 2 states, it is extremely important to quantify the expected value and express it in monetary terms, as the cost of implementation will inevitably be expressed in those terms.

In addition to the definition of one or more strategic objectives for each candidate application, the ways in which the achievement of these objectives will later be measured must be defined. If the strategic objectives can be quantified and expressed as a value, or a value over time, then it should be quite easy to measure the actual results of the application in monetary terms and compare them against the objectives. Objectives which are difficult to quantify may also be difficult to measure, but it is always important to do so. Otherwise evaluations after implementation of the application may be subjective and thus extremely difficult to put into perspective.

Some method of evaluation must be agreed upon. For example, if the objective is to provide new types of information to management, the managers involved might be asked to fill out a questionnaire evaluating the worth of the new or changed information after it has been available to them for 3 months. If the objective is to lower costs and/or to improve productivity, a value can be computed for the expected savings or productivity increase. After application implementation is complete, current costs can be measured for a period of time and compared to the objective. If the expected improvements concern productivity, then how much output each worker is producing during a stated period of time must be determined. From this information it will be possible to compute the workers' productivity (cost per unit of output), and compare it to the previous rate and also to the objective.

Objectives which deal with better service to customers or clients are among the most difficult to quantify and measure. One possibility is to define objectives in terms of how quickly certain functions ought to be carried out, for example, responding to a customer inquiry or purchase request. If this can be done, measurement after the new application is implemented will be quite simple. It may not be possible in a situation such as this example to translate this improved efficiency into monetary units—especially in a nonprofit organization. However, if this is impractical, objectives stated in terms of quicker service (or some similar criteria) are quite acceptable. The major task is to measure as objectively as possible how well the application goals are achieved.

Strategic objectives, no matter how they are expressed, as well as the methods and criteria for measuring the achievement of those objectives, must be defined by the line management of the organization—of course, in consultation with the system-design staff. The appropriate organizational level of management will vary, depending upon the scope and size of the proposed application. If a new application will serve only one organizational segment, the manager of that segment can define the objectives and measurement criteria. If a new application or set of applications will serve a wide range of organizational elements, all levels of management involved must cooperate in the definition of objectives and measurement methods. The broader the scope of the candidate application, the higher the level of management whose involvement will be essential.

This initial set of strategic objectives must be formally documented and agreed to, in writing, by the line manager(s) involved and by the system designers and their management. It is always appropriate to view each set of objectives as subject to change during system analysis, but in general the strategic objectives ought to remain stable. It may, however, be possible to further quantify both the objectives and the measurement criteria during the system study. Each change must be reflected in a revised statement of the objectives and formally agreed to by the parties who negotiated the original agreement.

IMPLEMENTATION-ORIENTED OBJECTIVES

Implementation-oriented objectives, while still stated as user requirements, are defined in computer-system-specific terms such as response time, system availability, and so on. The implementation-specific objectives must support the strategic objectives and can usually be related directly to them. For example, if a strategic goal is to improve the productivity of the personnel who handle reservations for a hotel chain, implementation-specific objectives must be defined that will provide a given set of functions, with a stated response time for reservation requests and a required level of system availability. Usually these objectives cannot be fully identified in the initial stages of system analysis and design, but they ought to be considered at this time. The items which make up the implementation-oriented objectives follow:

- Functions to be provided
- Response speed needed
- Availability requirements
- Integrity, security, privacy, and auditability specifications
- Degree of flexibility for change required
- Acceptable cost of implementation and operation (based on expected ROI)
- Schedule for beginning operation

Each of these objectives ought to be quantified, if practical, and methods of measurement must be defined to determine if the objective has been met. Four objectives—response speed, availability, cost, and schedule—can only be expressed quantitatively, and it is therefore relatively easy to define measurement methods and criteria. Functional objectives are not readily quantifiable, but it is usually possible to design tests which ensure that all functions are present and operate correctly. The remaining objectives are much more difficult both to quantify and to measure, but as they are extremely important to the success of the system, they require special attention. This topic is discussed again, in Section 3, at the point in data collection when implementation-specific objectives are defined in more detail.

If a first cut at specific objectives is made at this time, both line management and the system-design staff must be involved. It may also be appropriate to include some of the individual users in this process, as an introductory phase leading to data collection (Chapter 4). There is no need to have any formal agreement on specific objectives at this time, as they are preliminary and will certainly be changed as system analysis progresses. After data collection, the application objectives will be defined in more detail (Chapter 6), and at that time formal agreement is appropriate.

DATA COLLECTION

4

METHODS OF
DATA COLLECTION

After the procedures outlined in Chapters 2 and 3 have been carried out, one application will have been selected for system analysis and design. (The idea of analyzing one application at a time may be unrealistic but simplifies this discussion considerably. If multiple applications are being worked on simultaneously, each can be handled as described in this and the following chapters.)

The system-analysis process begins with a data-collection phase (step 4 of the process described in Chapter 1). The purpose of data collection is to accumulate enough information about the proposed application and its users to further refine the statement of system objectives (begun in the procedures described in Chapters 2 and 3), and to begin the process of system design (Sections 4 and 5).

The data elements which must be collected follow:

- The prospective users of the system

- The information needs of those users—that is, the types of output and output schedules required

- How the system will be used—interactively, in batch mode, etc.

- The processing required to produce the defined outputs

- The geographical range of the system—where the users are located and locations where computing equipment can be installed

- The data to be stored by the system

- Requirements for system integrity, security, privacy, and auditability

- Types of change and/or expansion likely to be needed (the degree of flexibility required)
- The managerial-control philosophy of the organization, as this affects the application

This may seem to be a forbiddingly long list of data to collect before system design begins. However, the design process—like computer processing—is very sensitive to the condition referred to in the adage "garbage in, garbage out." Inadequate data collection can invalidate an entire design, causing the system to fail when implemented.

It may of course be argued that some data elements simply do not exist. This is often the case when quantitative data such as transaction volumes or future numbers of users and terminals are needed. Many systems today, whether manual or computer-aided, are poorly instrumented; that is, quantitative measurements of work done are not collected routinely and accurately. In these cases two alternatives exist: either procedures must be instituted to accumulate the necessary statistics, which of course will require additional time during data collection, or estimates must be used. There is nothing inherently wrong in using data which may not be completely accurate; realistically, almost all measurements are subject to change and are therefore only accurate at a certain point in time. However, the system designer must clearly document the probable accuracy of collected data. If a number of the data elements used as input to the system-design process are estimates, or of suspect accuracy, special efforts must be made during system design to ensure the flexibility that will allow the system to adapt to change.

RESPONSIBILITIES OF USER MANAGEMENT

The data-collection process provides an excellent opportunity to establish good relationships with the people who will be the clients of the system and to enlist their support in its design, implementation, and operation. Their support can be a fundamental element in the system's success. An unhappy user can sabotage a computer-based system in subtle ways—sometimes simply by constantly complaining that it does not work correctly or is inconvenient to use. While the "unhappy user" syndrome can never be completely eliminated, it can be minimized. The time to begin this process is during data collection.

Ensuring good relationships between system users and system designers (and later with the operations staff supporting the system when in production) is a joint responsibility of the system-design staff and the management to whom the users report. The responsibilities of user management include the following:

• User management must explain the purpose of the new system to the users, emphasizing its prospective value to the users and the organization.

• They must explain the users' role in helping to define the system requirements, the importance of this role, and the procedure which will be used during the data-collection process. The actual definition of the procedure for data collection is the responsibility of the system designers, but it is more suitable for user management to explain the procedure to the users—at least in overview form. This will help to reinforce management's role as the sponsors of the new application.

• User management ought to introduce the system-design staff to the users and explain how they will work together in defining the system.

• Finally, user management must provide the necessary time for users to participate in data collection and the definition of system requirements. Users must not be asked to assist the system designers as a spare-time activity in addition to their full-time job; this will cause resentment and get the entire project off to a bad start.

The level of user management which ought to be involved will vary in different circumstances. In general, it is preferable for the users' immediate supervisor(s) or manager(s) to explain system goals and to sponsor the system-design staff. In some cases it may be useful to involve higher levels of management also. For example, if the system is expected to cause major changes in the way the organization operates, it will be appropriate for a high-level manager to explain the reasons for the changes and state how any resulting changes in the job position of any individuals will be handled.

An important task for management, with the aid of the sytem design staff, is to analyze possible negative reactions from the users and respond to these "up front" as far as possible. For example, the proposed introduction of computer-based capabilities almost always leads to rumors that people will be laid off, or that the system is so complicated that some people won't be able to learn how to use it and will lose their jobs, and other, similar fears. Facing rumors of this kind as early as possible is important. It is never possible to completely eliminate uneasiness and rumors when change is imminent; however, careful and continuing management attention to the problem can minimize its impact.

TASKS OF THE SYSTEM-DESIGN STAFF

The system-design staff also has a specific set of responsibilities:

• System designers must define an orderly method of data collection, with as little disruption as possible of the users' regular work routine.

• System designers must develop a high level of skill in personal relationships. This includes never using "DP jargon" when talking to the users, as well as avoiding any tendency to pose as the "computer experts" and look down on the users. The users are, in fact, the experts, whose knowledge of how the organization operates and accomplishes the necessary work is indispensable to the success of the computer-aided system.

The mechanisms used for data collection will vary, depending upon the size and complexity of the system being studied, the number of users, and the type of work they do. One method which has been used successfully consists of the following three steps:

1. The system designers explain to the users the purpose of the data-collection process and describe what kind of information is needed.

2. A set of data-collection forms (see example forms in the Appendix) is left with each user. The users are asked not to fill out the forms at this time, but to look them over and think about how they might fill them out.

3. After a few days (not less than 3 or 4 days, but preferably not more than 2 weeks), the designers begin defining system requirements with the users, and the data-collection forms are filled out as discussions progress.

This procedure removes the pressure on users to respond immediately to questions for which they may not be prepared. It also avoids the difficulties which may be encountered if users are asked to fill out data-collection forms and turn them in without any interaction with the system design staff. Ambiguous responses to questionnaires are common, and even responses which do not seem to be ambiguous may not be accurate. If a system designer fills out each data-collection form as part of an interview with one or more users, the entries will generally be accurate and the system designers will also gain a more in-depth knowledge of the users and their work.

If a very large number of users are involved, it may be impractical to interview each individually. However, at least some users ought to be interviewed in depth. Consulting with a group of several users and using the information obtained to fill out one set of data-collection forms is also a useful means of obtaining accurate data. Regardless of which method of data collection is used, it is important to involve as many individual users as is practical. It is *never* adequate to interview only supervisors and/or only one or two users. Even in supposedly homogeneous groups doing the same job, there are often different methods and different viewpoints which can shed important light on system requirements. In addition, talking with as many users as possible can help ensure that all feel they are part of the system-definition process—and that the system is not being imposed on them.

IDENTIFYING SYSTEM USERS

Because the data-collection process requires close cooperation between the system designers and the prospective users of the system, the first step in data collection is to define who the system users will be. In some cases it is very easy to identify the users; for example, in an on-line-banking system, bank tellers will be the primary users.

In other cases it may be much more difficult to decide who the users will be, especially if the definition of system objectives is general. For example, it may have been determined that the cost of processing the orders received for a company's products is too high, lowering profitability. A computer-based system to handle the entry and processing of orders may seem to offer a solution to this problem. However, it might be unclear initially whether sales personnel ought to be asked to enter orders directly via terminals or whether a central order-processing staff using the terminals would be more efficient and cost-effective. It would be important, in this example case, to identify the sales staff, the order-processing staff, and possibly even the company's customers (who might be asked to enter their own orders) as potential users of the new system.

Identifying potential system users is therefore a typical system-analysis problem. It will be necessary to study how the required functions are provided at present, who provides them, and the job focus of each person or group of people involved. A form such as that shown in Figure 4-1 may be useful in recording this information. This form is less detailed than would be used in work analysis for purposes such as time-and-motion studies. The main emphasis here is to determine work flow, data elements, and necessary functions in the existing system.

To reiterate, it is important to involve as many individuals as possible in the data-collection process. However, in describing users it is generally possible to group individuals into related categories. For example, when

FUNCTION: PROCESSING FLIGHT RESERVATIONS	
Operations	*Data*

1. Answer telephone.
  a. Record data.
	Desired flight number *or* origin
	   and destination cities and
	   desired time of departure or
	   arrival
	Date of travel
	Class of accommodation
	Number of persons

2. Determine flight number if not given.	Flight schedules, arranged by origin, destination, and time (both departure and arrival)
3. Determine flight availability.	Booking data for specific flight and date
4. Inform customer of flight status.	
5. Book flight if available and customer wishes to do so.	Customer name, telephone number
6. Determine method of payment.	Credit card type, number, expiration date, customer address if appropriate
7. Confirm reservation to customer. a. Break telephone connection.	

Figure 4-1 Data elements and work flow.

analyzing a company's sales force it may be appropriate to group people as follows:

- Wholesale sales representatives
  —High-volume items
  —Low-volume items
- Retail sales representatives
  —Direct customer sales
  —Sales to resellers
- Management of the sales organization

These categories are of course only examples, and are not necessarily appropriate for any particular sales organization. The important thing is to recognize the characteristics of certain users' requirements which set them apart from those of other users, and then to classify users with similar needs into cohesive groups. In the above example, salespeople handling high-volume items may make multiple sales each day. Those who handle low-volume items, perhaps with a high value per item, may spend days, weeks, or even months making a single sale. The data-input rates and output requirements of the two groups will therefore be quite different.

Any experienced system designer will recognize that it is impossible to avoid doing a certain amount of broad-brush system design during the data-collection process—and even while identifying system users. System design is not a neat, tidy process in which each step is independent of other steps and proceeds in a mathematical sequence. If that were the case, computer programs might well replace system designers. In fact, data collection is part of an interactive, iterative process leading to a finished design. Designers will invariably form trial designs during data collection; often it is impossible to conduct meaningful interviews with potential users without doing so. However, the tendency to think of one trial design as the final design and to bias all data collection in favor of that approach, is a dangerous pitfall which must be avoided. It is essential that the system designers retain open minds during the data-collection and data-analysis phases, so that the design which is finally chosen rests upon an objective understanding of the users' requirements.

In addition to an understanding of each user group's functions, it is necessary to determine and document each group's level of experience with computers and general level of education. For example, a situation in which the data-collection process is related to the improvement and expansion of an already-computerized application in which most of the people in the organization use terminals is quite different from one in which data collection is for a new system in an organization in which the prospective users have never used terminals or even computer printouts. Still another situation exists if a new system is being considered in a college where all the students are required to take at least an introductory computer science course. In this case a system designed for student use could assume not only familiarity with terminals but a moderately high degree of data-processing knowledge.

Although questions about users' educational level can be quite sensitive, this information is very important. A highly educated group of professionals—such as lawyers, doctors, executives, or nurses—will be offended by simplistic or casual system interfaces which may be appropriate for other groups. For example, the greeting "Hi!" displayed on a terminal screen may be reassuring to a factory worker but demeaning to a high-level manager. Finding the correct level of computer interaction for each group of users depends upon understanding that group's knowledge and cultural background.

For each group of possible system users (some of whom may not actually be users when the system design is complete), descriptive data collected must include the following:

- User-group name (any descriptive title, needed only for convenience in referring to the group).

- Functions related to the proposed system which are performed by this group. Some indication is required of how many of the functions are routine and can be completely predefined and how many are in response to unusual circumstances and are therefore ad hoc in nature. In general, the functions listed ought to be those described in the documentation of data and work flow (see Figure 4-1).

- Number of people currently in the group.

- Level of collective experience with computers and terminals. Some individuals within the group may have substantially more or less experience, but the system must be designed for the "typical" user, insofar as such a person can be described.

- Degree and frequency of personnel turnover within the group.

- An indication of the workload of the group. Are they always extremely busy, with more work than they can comfortably handle, or is there plenty of time for breaks, chatting, and so on?

- General educational level within the group.

- Any other useful data which can help to better describe the group, its functions, and its requirements.

Figure 4-2 shows an example of how these collected data might be documented. Note that this example describes the user group whose major functions are documented in the example in Figure 4-1.

MACHINE USERS

This discussion has been oriented toward human users, but machines may also supply input to, and/or accept output from, the new system. For example, an integrated factory-control system may have to be connected to existing computer-controlled machinery. In a hospital or research environment an integrated information-processing system may require connection to patient-monitoring or laboratory equipment. A new system may also require connection to existing information-

USER GROUP: RESERVATION CLERKS	
Functions performed	Processing flight reservations
Number in group	48 clerks (24 on day shift, 24 on the two night shifts), plus 5 supervisors (who can act as clerks in emergency situations)
Computer experience	All are experienced with the use of hard-copy terminals; some have used video display terminals
Personnel turnover	Currently 12 percent per year. Over 5 years has ranged between 8 and 15 percent.
Workload	Varies considerably during the 24 hours/day, 7 days/week schedule, with quite well-defined peak periods. There are also well-defined seasonal peaks and peaks associated with holiday travel.
Educational level	Predominantly high school graduates.
Other information	The group is highly motivated, and very willing to participate in the definition of a more advanced computer-based system.

Figure 4-2   User-group description.

processing system(s) used by the same organization, and/or to computers operated by the suppliers, customers, or clients of the organization.

For each machine which will potentially be connected to the system, the following data are required:

- Machine-type name (equivalent to user-group name)

- Functions performed by the machine

- Input used by the machine and method(s) by which data are currently supplied

- Output and output methods currently in use

- Capabilities (additional functions) of the machine which are not used at present

- Level of programmability—is this a computer-based, programmable device or hard-wired?

- Any other data useful in describing the machine and its capabilities

MACHINE TYPE: EQUIPMENT-SCHEDULING SYSTEM	
Functions performed	Determines how and when each aircraft will be used, supplies (meals, etc.) required, crew requirements.
Data used	The equipment-scheduling system supplies flight capacity numbers to the reservation system, and receives passenger load data from it.
Input/output	A high-speed communications link is used to exchange data between the two systems.
Additional capabilities	The system is implemented on a large-scale general-purpose computer, so that numerous features could be added.
Level of programmability	See above.
Other information	Complete documentation of the equipment-scheduling system is available.

Figure 4-3   Connected machine description.

Figure 4-3 shows how data describing machine "users" of a system might be documented. This example is related to Figures 4-1 and 4-2 and is somewhat simplified for reasons of space. However, it is unnecessary to completely describe computers or other machines which will be connected to the system being analyzed; only facts relevant to that system are required.

Figures 4-2 and 4-3 are only suggestions for documenting the descriptions of system users. Any suitable format can be used; it is, however, extremely important to document these findings formally. It is also important that each designer involved in data collection document his or her findings consistently. This will ensure easy communication between the designers of the system and reduce the vulnerability of the design process to personnel changes in the system design staff.

The process described in this section should be sufficient to identify the people and machines which may interface with the new application and which can therefore be viewed—in a broad sense—as its users. In a complex system, a more complete description of the users will be obtained during the data-collection process. The most important goal at this stage in system analysis is to identify all the prospective users, so that their requirements can be included in the analysis and design of the new system.

5

DATA TO BE
COLLECTED

Once the users of the system have been tentatively identified, the data-collection process can begin. The data elements to be collected follow:

- Output requirements
- Usage modes
- Processing requirements
- Geographical locations
- Data-storage requirements
- Integrity (including correctness and availability), security, privacy, and auditability requirements
- Expected change
- Managerial control

The purpose of data collection is to complete the detailed statement of system requirements and objectives, as described in Chapter 6, and to obtain the information necessary for the system-design process, as described in Chapters 7 through 13. Data collection, like other facets of system analysis and design, is an iterative process, but of course the goal ought to be to obtain as much information as possible during the initial interviews with prospective users.

The data-collection methods described in Chapter 4 must be tailored to the specific users and environment. In general, the system designers can use a set of data-collection forms, similar to those provided in the Appendix, as the basis for discussions with the users—either individually or in groups. If the users are given an opportunity to study the forms in advance of the interviews, it will speed up the process as well as improve the quality of the users' contributions. This chapter describes the data elements which must be collected and documented before system design can begin.

OUTPUT REQUIREMENTS

One of the most important aspects of a new system is the definition of output-information requirements. Each information system can be viewed as having one major goal, which is to provide the information necessary to assist its users in the execution of their tasks. Output needs can be defined by answering the following questions:

1. What information is required to carry out each of the functions of each of the user groups (as identified during the system study discussed in Chapter 4)?

2. How rapidly, or how often, is each output response or report required?

It is unnecessary to define each output in detail during the data-collection phase; it is also unnecessary to design specific output or report formats. However, it *is* important to determine the major output data entities, as this will affect the requirements for input data as well as the design of the database(s) which support the system. In an on-line order-entry application, for example, it is important to know whether prices are needed on the order-acknowledgement documents which are provided to the order-entry personnel, to the customer, to the salesperson, and to the personnel at the warehouse. Probably some of these output documents or displays must include the price for each item ordered and others need not. This type of information must be obtained during data collection.

Output requirements can be grouped as follows:

• Responses to interactive transactions entered routinely

• Periodic, event-triggered, or on-request reports which can be predefined

• Responses and/or reports for which the exact content and frequency cannot be predefined, e.g., ad hoc reports

Not every system requires all three types of output, but all must be considered. Special attention must be given to output which cannot be predefined in detail but which generally can at least be categorized. This type of output may dictate that specific data elements be included in the system database and may also indicate the need for ad hoc query and/or reporting software. (The capability for ad hoc access to stored data is becoming a requirement in all computer-based systems.)

It is necessary to define how often each type of output will be required and how rapidly it must be provided after the request for output is made. Common terminology for the speed of output preparation follows.

- For interactive processing, *response time requirements* indicate how rapidly an output response is expected to follow the input request.

- For some batch-processing operations, *turnaround speed* indicates how quickly the output is expected after the job or the request for the report is submitted.

- For other types of batch processing, as well as for certain types of interactive reports, *due date and time* are used to define when a regularly scheduled output must be produced.

During system design and implementation, emphasis tends to be on meeting interactive-response needs. However, in most systems it is equally important that other output requirements be met—especially in the case of date and time requirements. Usually there are also some classes of work which have a fairly wide latitude in scheduling, and these must be identified during data collection.

Realistic output-speed requirements *must* be set. Users may tend to define unrealistically rapid response times, while system designers may underestimate the importance of speed. A designer may, for example, believe that 15-second response is just as good as 10-second response, since 5 seconds is a relatively short period and 15-second response may be much more economical in terms of computer resources. The required response speed is often closely tied to the working environment of the users. An airline reservation clerk handling hundreds of customer inquiries each hour needs very fast response, in the range of 1 or 2 seconds. A teller working with customers in a bank may need only 3- to 5-second response, as the number of customers to be served is smaller and the need for quick response may be less urgent.

It is almost always useful during data collection to provide the users with prototypes of input and output to help clarify both the output-data requirements and the response-speed requirements. If the users of the new system are experienced with terminals, it may be relatively easy for them to define their future data needs and response requirements. In contrast, people with no prior experience in terminal use may find it difficult to visualize this form of interaction. Helping them to do so will generally improve the quality of the users' contribution to data collection—and minimize the possibility that user requirements are inadequately or incorrectly defined.

It is unnecessary to build a prototype of all (or even any large part) of the processing system at this stage. It is, however, useful to design some trial input and output formats for the major types of transactions. It will probably be necessary to write a small driver application and to create a small file or database with sample data. For example, for an on-line-banking application, it would be appropriate to design the terminal-screen format(s) for customer deposits and withdrawals, write a small program to manipulate those screens, and create a small file of fictitious customer accounts. This would allow the tellers to visualize how the terminal-based procedures could fit into their normal operation.

The value of prototype interfaces cannot be overestimated. Too often, however, the time and potential expense involved make organizations reluctant to create prototypes. If, for example, there are no unused terminals available, it may be necessary to lease or purchase some for the prototype phase. This may be particularly troubling to management if the vendor from whom terminals will be procured has not been selected, as it is possible that devices purchased for the prototype phase cannot later be used in the production system. While these are legitimate concerns, they must be weighed against the total cost of system design and implementation and against the risk of implementing an unworkable system because of an inadequate understanding of user needs. Again, the focus of this book is on complex system, especially distributed systems. These systems represent major investments, and in the context of such systems, the cost of several terminals for the development of prototype interfaces is minimal.

A small personal computer, such as an Apple or TRS-80, can also be used as the base for prototype development. These systems are inexpensive enough to be dedicated to building prototypes, especially in a large system staff. If this approach is chosen, however, care must be taken that the prototype interfaces do not differ dramatically from the interfaces which will be provided by the production system. Some personal computers, for example, feature color displays. It would be unwise to use a color display during the prototype phase if the terminals to be used later provide only black-and-white screens; the terminal users might become accustomed to color, and react negatively to its absence. It is also important that a prototype avoid both extremely fast and extremely slow response. If users first interface with a prototype which provides responses in less than 1 second, they may be disappointed with anything significantly slower. Since many production systems do not really require 1-second response, it is important not to raise unrealistic expectations.

Initially the response, turnaround, and due date and time requirements can be defined in a relatively general way. For example, interactive queries might be defined as requiring 3-second response. As system analysis progresses, this must be refined to indicate the range of acceptable responses and the number of transactions which must fall within that range. Before the users and their management are asked for formal concurrence in the system definition (see Chapter 6), the 3-second response requirement might be restated as:

- 95 percent of all transaction responses of type x must fall within the range of 3 to 5 seconds.

- All transaction responses must occur in less than 10 seconds.

One of the most common problems in on-line systems is inconsistent response speed. Response which is too fast in the user's opinion may be as disruptive as that which is too slow. Within reasonable limits, it is better to design for slower but consistent response speed than for faster, less consistent, speed. If the user knows what to expect from the system, he or she will be comfortable using it. If the system is unpredictable in its speed or actions, the user will be uncomfortable and distrustful. Response speed is therefore one of the aspects of system design which requires a great deal of careful attention.

The goal of output definition is to identify each major type of output needed. Realistically, not every type of output can be defined at this time; that is one of the reasons that a flexible, expandable design is required. However, all the main-line responses and reports must be described. For each of these, the following information must be collected and documented:

- Output name

- Data elements required in this output (e.g., paychecks require the data elements *name, check amount, date,* and possibly *employee number* or *social security number.*)

- Input or event trigger which causes this output to be produced (e.g., a query response is produced because of a query input; monthly reports are produced because monthly activities are complete and some specific trigger event is selected to indicate that completion; batch output is produced because the batch job is entered, etc.)

- Response speed, turnaround time, or due date and time required

- User group which will receive this output (identified by the user-group name assigned earlier; see Chapter 4)

- Any other necessary information

Figure 5-1 provides an example form for the documentation of output requirements. As in all phases of data collection, the output descriptions documented at this time may be changed as data collection proceeds and possibly also during data analysis (Section 4) and system design (Section 5).

USAGE MODES

Each system will probably involve multiple modes of use, although interactive processing now dominates batch-mode use in new applications. The possible usage methods are as follows.

Interactive, predefined operations, such as the input of certain types of inquiry and/or update transactions followed by output responses, are most used in production applications which require rapid responses. An

OUTPUT: WEEKLY SALES SUMMARY (ID: WSS)	
Data elements	1. Department number, item number, item description, sales volume, variance—for *only* items which vary from 6-week rolling average by ±15 percent 2. Same information, on request, for all items regardless of variance
Trigger	1. Completion of on-line business each Saturday 2. On request, via interactive transaction
Response, turnaround, due date/time	1. At start of business each Monday 2. Within 4 hours after request
User group	Department buyers (Group ID: BYRS)
Other information	Normally the report is required as printed output, with the section for each department delivered to that department's buyer(s). A buyer may select an option to have that department's part of the report retained on-line, for selective scanning from a terminal.

Figure 5-1 Output requirements.

on-line order-entry application fits into this category and might be an improvement over a system in which orders were recorded on typed documents, then later key-entered (to disk, tape, etc.) and processed in batch mode. Interactive processing would allow immediate order (or back-order) confirmation to customers and, compared to the batch system, might also represent a reduction in the cost of handling hard-copy order documents.

Interactive ad hoc queries and/or reports are necessary when it is impractical to predefine all types of output required. Systems which serve professional or managerial users are likely to need ad hoc capabilities because those users often have one-time data requirements which can most effectively be met by a generalized and easy-to-use query system. For example, when interest rates are high, many companies slow down their payment of bills. An accounts-receivable manager might want to determine the trends in the promptness of payment by major customers or by all customers. This ad hoc report would be required even in a system designed to automatically report delinquent accounts.

Although the use of ad hoc queries and report requests cannot be exactly predicted or exactly quantified, it is extremely important to determine as accurately as possible how many users will need these capabilities and how often they will use them. Because of the unpredictable nature of ad hoc facilities and the amount of database searching necessary to answer some queries, this mode of operation can be a heavy load on system resources. In addition, an easy-to-use query capability will bring into play the "turnpike effect" well known to designers of highways. When it is difficult to travel between City A and City B because of poor roads, only people who have an urgent need will travel that route. However, if a turnpike is built between the cities, but with only enough lanes to handle the previous traffic, it will typically become overloaded almost immediately. The ease of travel on the turnpike has in effect generated new traffic. Similarly, when it is hard to obtain data from a computer-based system, many people abandon the effort, leading to the erroneous impression that few or no users have data-access requirements. However, if it becomes quick and easy to access data, many users who have legitimate requirements suddenly surface. The definition of requirements for ad hoc facilities must therefore be approached with considerable caution.

Regularly scheduled batch processing is often used to produce periodic reports or output, for example, payroll checks. While it is generally true that new systems and new applications include less batch processing as a proportion of total processing than older systems, there will always be a certain number of reports which are most efficiently and economically produced in this way. This category may also include the preparation of

---

APPLICATION: RETAIL SALES ORDER PROCESSING

Interactive, predefined operations

1. Point-of-sale transactions
   a. Sales
   b. Returns
   c. Adjustments
2. Pricing
   a. New items
   b. Price adjustments
   c. Time-specific sale pricing (e.g., 15% price reduction)
3. Customer accounts
   a. Establishment
   b. Change
   c. Cancellation

Interactive, ad hoc queries and reports

1. Sales history, selected item(s) (Report ID: SH)
2. Purchasing status, selected item(s) (ID: PS)
3. Customer credit account status (ID: CAS)
4. Supplier account status (ID: SAS)

Regularly scheduled batch processing

1. Daily sales totals (ID: DST)
2. Weekly sales summary (ID: WSS)
3. Accounts payable (ID: AP)
4. Customer statements (ID: CS)

On-demand batch processing

1. Sales summary (ID: DSS)
2. Customer account trends (ID: DCT)

Periodic on-line reports

1. Continuous system and network status display (ID: ADM1)
2. Notice of error, failure, unusual system load (ID: ADM2)

Program preparation

Used by IRM staff only

Figure 5-2 Usage modes.

output for archival storage, for example, microfilm records created to meet a legal requirement for the long-term storage of transaction records.

On-demand batch processing includes jobs not predefined as part of the production workload but submitted when required by system users. This category can be expected to become smaller over time, especially as ad hoc data-access facilities come into wider use.

Periodic or event-driven on-line reports include system-monitoring capabilities required by the administrators, for example, continuous or periodic displays of the activities in process and the status of the communications network. This category may also include notices of unusual events sent automatically to specific administrators, for example, a notice to an auditor that a transaction involving an unusually large amount of money has just been initiated or a notice to a security administrator that an attempted security breach has occurred. Output of this type may be displayed but is usually also printed as a long-term log—either immediately, or periodically after accumulation on disk. Adequately defining the requirements for reports in this category reduces the need for regularly scheduled batch reports.

Program preparation (compiling, debugging, etc.) to support the staff which implements and maintains the system, is of course always required. In some cases, non-data-processing users will also require program-preparation facilities—either using programming languages such as BASIC or Fortran, or as extensions to ad hoc facilities to allow the preparation of formatted reports with column totals and similar features.

Most of the modes of usage which will be needed in the system will be documented as part of the definition of output requirements, as described earlier in this chapter. If necessary, a summary of usage modes can be prepared, as shown in Figure 5-2. If used, this type of summary must be cross-referenced to, and consistent with, the documentation of output requirements. It should also be noted that the analysis of usage methods may lead to the definition of output requirements not uncovered earlier. Those new requirements must be added to the documentation created earlier in the data-collection process.

PROCESSING REQUIREMENTS

Data collection also includes the definition of the processing which must be done by the new system. Specific application *programs* are not defined at this point, but rather the needed application *functions*. This definition must include a general statement of what processing is to be done. For example, one required function in an on-line order-entry system is the processing of new orders. This processing might be defined as shown in Figure 5-3.

This example shows that only major functions are documented. In the example step 1, verifying customer identity, causes a decision which may lead to continued processing of the order or, if the customer-identity number is invalid, to the generation of an input-error notice. Other decision consequences are similarly documented only in the form of the

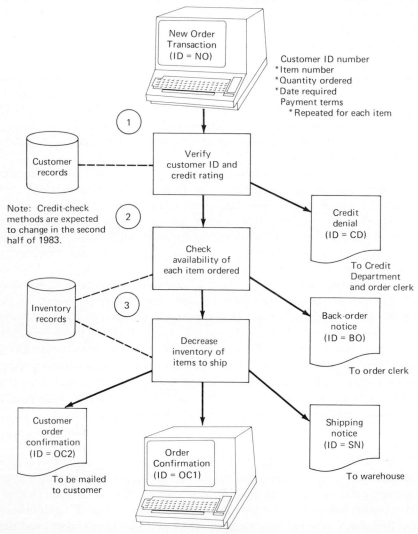

Figure 5-3 Processing flow. Process: On-line order entry, new orders.

different types of output produced. If the processing functions are extremely complex, it may be useful to document them in more detail than shown in the example. In fact, a top-level flowchart may be appropriate in those cases. The main reason that a typical flowchart form is not used in Figure 5-3 is that this document will be used to confirm the processing needs with the users. This simplified-flow form is generally easier to read than a flowchart. However, any format or method which is mutually agreed upon by the designers and users of the system is appropriate.

Note that Figure 5-3 shows several types of output. If any of these is a new output, not defined earlier, it must be documented. Some designers may, in fact, prefer to define and document processing requirements before documenting output requirements. The author's preference is to treat output first, but any sequence that an individual designer or design team is comfortable with is acceptable.

If the system is very complex, with a correspondingly large number of input, output, and processing functions to be described, it may be advisable to assign an identifier to each input and output, as shown in Figure 5-3. Identifiers allow easier and more explicit cross-referencing among the sheets used to document output requirements, processing requirements, data-storage requirements, system users, etc. These cross-references are used extensively during data analysis (see Section 4).

GEOGRAPHICAL LOCATIONS

The geographical locations involved in the application must be determined and documented. This information is most directly related to the design of the communications network (Chapter 12), including the decisions on where terminal devices will be located. If a decentralized or distributed structure is chosen (Chapter 10), the information concerning location will also help determine where computer equipment is to be placed.

The locations of all the potential users of the system must be documented, as shown in Figure 5-4. City, building, and room locations are all needed. Many new systems will use some form of local-area network (LAN) to link terminals and processors which are within a building, or perhaps in adjacent buildings. (LANs can cover extensive areas; there is at least one Xerox Ethernet network which extends nearly 1400 meters.) It is not necessary at this time to obtain measurements of distances between user locations and existing or potential computer sites or to convert city locations to vertical/horizontal (V/H) coordinate form.* That

---

* V/H coordinate form is the location system used in the U.S. telephone network to precisely locate any site on a digitized map of the country.

SITE: SALES-ORDER-PROCESSING CENTER	
Location	Headquarters building, Nashville, Tenn., fifth floor (entire floor is occupied by the sales-order-processing staff)
User group(s)	1. Sales-order clerks 2. Sales-department supervisors
Machine(s)	None
Environmental conditions	Office, temperature-controlled between 19 and 25°C
Other information	Planned expansion of the sales-order-processing staff includes use of the fourth floor of the same building; environmental conditions identical with fifth floor

Figure 5-4 Geographical locations.

more detailed information will be needed later, during system design (Section 5). However, during data collection the definition of system users may still be inexact—potential user groups may not, in the final system design, have any contact with the computer-aided system. It may therefore be a waste of time to collect detailed data, such as distances, which will not be needed.

The entries shown in the example are generally self-explanatory. For each site, the user(s) and/or equipment (computers, other machines) located there must be documented. The user-group and machine names assigned in Chapter 4 must be used to provide a cross-reference among users, equipment, and locations.

The environmental conditions at each location must also be defined. This information will be required when decisions are made concerning where distributed or decentralized computers can be installed. Only a general indication of the environment is required at this time; possible conditions include the following:

- Temperature- and humidity-controlled (e.g., computer room)
- Office
- Factory, normal
- Factory, extreme
- Out of doors

The category "factory, extreme" covers situations in which a corrosive atmosphere, extreme heat or cold or humidity, or vibration (perhaps caused by the operation of heavy equipment) makes it impractical to install computer equipment unless it is specifically designed to operate under those conditions, or *ruggedized*. Ruggedized computers are most used by the military, for applications such as shipboard systems, battlefield support, and airborne systems. However, as computer-aided systems move into many new areas, ruggedizing may be required in areas other than defense and aerospace.

One other point to be considered is whether any users will require terminal access to the system while traveling. Sales representatives, for example, might submit orders directly from customer sites. In an electronic-mail facility, users will want to access their messages even when on business travel out of town or perhaps even out of the country. If the system must support requirements such as these, the necessary information must be documented with the other location data.

DATA-STORAGE REQUIREMENTS

Another step in data collection is to define the data which the system must store on a semipermanent basis. This is not the same as designing the system database(s) or files; it simply involves defining the data elements to be included in those files and/or databases. The process of defining data to be stored will naturally provide an indication of how data elements must be grouped to meet system needs. However, as in other aspects of data collection, it is important to resist the tendency to begin system design before all data have been collected and analyzed.

The definition of output requirements is closely related to the definition of data-storage requirements. The definition of data requirements for the entire system really begins with output definition. Any required output data elements which are not provided as system input or the result of computation must be obtained from system storage.

This view of data relationships is somewhat at odds with the currently popular view of information systems as being data-centered. In that view, the first step in system design is to determine all the data requirements and data relationships of an organization and describe those requirements and relationships in terms of what ANSI (the American National Standards Institute) calls a conceptual schema, and then translate all or part of that conceptual schema to an actual database design and definition. Application systems can then be built around that database. The theory is that a thorough definition of all the organization's data relationships will provide the necessary stored data for all possible applications.

This is clearly an attractive and logical view of how to design information systems, whether fully computerized, completely manual, or computer-aided. However, it is rare that an organization has the time, the money, or the expert system- and process-analysis staff necessary to tackle this definition in a full-scale way. In general, therefore, treating system design as being output-driven and then using as much as is practical of the data-relationship approach is a useful compromise.

In this compromise approach, needed data elements can be defined and grouped as follows:

1. Those required to create an output response or report defined during data collection

2. Those required because of existing organizational data relationships

As an example of group 1, one required output in a factory-management system might be an inventory-status report, available on request, which would include the following data elements:

- Item-identifying number

- Item description

- Current stock balance

- Warehouse location(s) where stock is held

- Order status of any outstanding orders

Since none of the required report elements can be computed, all of them must be included in stored data supporting the system. It can be readily seen that any system will include many overlapping groups of data elements; in the above example every data element will probably appear in at least one other report or response, and some may appear in many.

Group 2 deals with the fundamental structure of the organization and how data elements are used in the functions which the organization performs. (This is the basic premise of the data-centered view of system design.) For example, an analysis of a factory-management system would define the need for item-source information. That is, each item used in the factory but not manufactured there would be purchased from one or more suppliers. Even though the computerized system might not be intended to include the generation of purchase orders for out-of-stock items, it is probable that this function would be added later. The necessary data elements to support that application ought to be defined, especially since they are so closely related to the inventory data.

The question of how far to carry the identification of data elements in group 2 is not an easy one to answer. Carried to an extreme, this could involve the identification of all data elements and relationships in the organization. If ignored completely, databases could be designed which would require major reorganization as soon as new applications were selected for implementation. Neither of these extremes is desirable. A good rule of thumb is that data elements closely related to those needed for the application(s) under study ought to be identified and documented.

In a point-of-sale system in a supermarket, for example, data elements related to items and their prices would be identified as part of group 1. Since the person operating the checkout equipment must usually be identified for security purposes, it is logical to consider the system as a source of hours worked (at the checkstand) per person. The data necessary to support labor cost and payroll applications, such as person identifier and hourly rate, should be documented. Later in the system-design process a decision can be made to include or exclude those elements in the system's database.

The data elements defined in both groups 1 and 2 must be documented; an example of this type of documentation is shown in Figure 5-5. For each data element the following definitions are required:

- Data-element name

- Number of bytes or bits required to store this data element

- Major use of the data element

- Logical grouping to which the data element belongs

- Integrity, security, privacy requirements

- Expected change, indicated only as "yes" or "no"

The *major-use* entry should include the identity of the most important response or report in which this data element is used. There are, of course, data elements which need no entry under major use, which is largely a means of justifying the inclusion of the element in a database or file. In a personnel application, for example, data elements such as employee number and name are self-explanatory, and require no justification. An unusual data element, such as a person's experience in fields unrelated to the organization's operation, would require justification via the major-use entry.

The *logical-grouping* entry is optional but may be useful as a first step in database design. Many applications require both on-line and batch-mode output. Data elements used in the on-line portion of the applica-

APPLICATION: FACTORY-FLOOR MANAGEMENT

Data element	Size	Major use	Logical grouping	Integrity, security, privacy	Change (Y/N)
Finished-item identifier	12 bytes		Inventory	Normal	N
Parts-item identifier	12 bytes	Bill of material	BOM	Normal	N
Item description	36 bytes	Output documents	Inventory, BOM	Normal	N
Machine identifier	6 bytes	Manufacturing-process control	Process definition	Normal	N
Process sequence*	Variable	Manufacturing-process control	Process definition	High integrity, secure	Y

* Step-by-step definition of a process; e.g., how to manufacture a specific part.

Figure 5-5 Data-element definitions.

tion will usually be stored in a database on rapid-access storage, while those elements used only for batch processing may be stored in files or on slower-access devices. Since the data-collection process provides much of the information necessary to make these distinctions, grouping data elements at this point can be very helpful. Some system designers may choose to organize the documentation of data-element requirements by logical groups rather than by data elements. Either sequence is acceptable, as long as an open mind is retained until data-storage design begins (Chapter 11).

The column labeled "integrity, security, privacy" allows protection requirements to be specified for each data element. These requirements are discussed later in this chapter. The column which indicates expected change is also discussed later.

PERSONAL-DATA-STORAGE REQUIREMENTS

This discussion has been focused on data to be stored in system-managed databases and files, but it is also necessary to define the requirements for temporary or personal files. This category does not include system-support files such as backup copies of database information, but rather files needed by the system users. Some systems will not include any storage of this type; however, more and more systems in the future will do so. One example occurs in the use of time sharing by professionals such as engineers. Each user of this type may require files in which to store not only data being worked on but also personal programs created by that user. Increasingly, managers, professionals, and their staffs will want to store a variety of data sets such as tickler files, personal schedules, often-used telephone numbers or addresses, and similar information.

Files or databases in this category cannot usually be defined in the same detail as system-managed data storage. However, a great deal of storage space could be required for these purposes, so it is essential to define the needs adequately. Figure 5-6 shows an example of how these personal-data-storage requirements might be documented. For each data set of this type, the following characteristics are required:

- Data-set (file) name

- Function for which used (program storage, personal schedule, etc.)

- Estimated amount of storage required

- Average length of time the data will be retained

DATA SET: PERSONAL SCHEDULE	
Function	Keep track of appointments, meetings, travel, etc., for one manager or professional
Amount of storage	50,000 bytes per month; 3 to 6 months on-line desirable
Retention time	Permanent; rotating use, with oldest month's data discarded each month
Mode(s) of use	Interative creation, access, and update
Access speed required	3 seconds for specific inquiry; 10 seconds for search (e.g., find open 3-hour segment during next 3 weeks)
Number of users	2 (manager or professional plus secretary)
Other information	Future application will allow schedule matching among up to 24 individual schedules, so that meetings can be coordinated

Figure 5-6 Personal-data requirements.

- Predominate mode of use, e.g., use of entire data set as input to a program, searching for a specific item using one of several possible keys, etc.

- Access speed typically required

- How many people will use the data set

- Any other useful information

System analysis and design, especially of large centralized systems, have not typically included the definition of files of this type, except in the case of time-sharing systems. However, two trends will emphasize the need for the storage of personal data. These trends are:

1. Increasing use of terminal-based systems in offices, often with direct terminal use by managers and professionals

2. Growing use of personal computers, so that many managers and professionals have some experience in computer use, and possibly in programming

Although these trends do not imply the demise of large shared databases—and in fact these trends will probably contribute to database growth as the need for large volumes of stored data intensifies—they do mean that a great deal of attention must be paid to the management of personal files or databases to provide user responsiveness. During the data-collection process the requirements for this type of data storage must be as fully defined and documented as possible.

INTEGRITY, SECURITY, PRIVACY, AND AUDITABILITY REQUIREMENTS

Although each of these four items is inherently important in system design, there are so many interrelationships among them that it is best to treat them together, both during data collection and later during system design (Chapter 13). *Integrity* deals with data correctness and with the ability of the system, or a part of the system, to continue or resume operation in spite of errors or failures. *Security* involves protecting the system and stored data from unauthorized access or use. *Privacy* is specifically concerned with the protection of personal data from improper disclosure or use. *Auditability* deals with the ability to examine the system and its output to verify that it is operating correctly within the relevant guidelines (and possibly statutes).

Every system will have requirements for integrity, security, and auditability, and many will also have privacy requirements. During data collection the system users, their management, the legal staff, the auditor's staff, and possibly other officials of the organization must be interviewed to determine exactly what these requirements are.

INTEGRITY REQUIREMENTS

Integrity requirements include two major elements:

- The degree of data correctness needed
- The level of system availability needed

Although 100 percent accuracy in data handling is desirable, it is never achieved, as the cost involved would be too great to be practical. All systems therefore tolerate some level of error in return for lower cost. During data collection the degree of error tolerance in the application under study must be determined. If the application is currently handled by manual processing, error checks and error-handling methods no doubt exist. These will usually provide an indication of which data elements are most important.

Emphasis in integrity protection is usually placed on financial data, or on quantitative data such as numbers of items ordered, shipped, or stocked. Even here, correctness is usually more important when large amounts of money are involved. An error of 10 percent may be acceptable on an amount of $10, but clearly unacceptable if the amount is $100,000. Similarly, a variance of 10 percent between the actual and recorded inventory levels may not be worth worrying about if the items stocked are metal washers worth $0.005 each; such a variance is much more serious if the items are expensive television sets.

Many organizations have either formal or de facto rules defining the level of correctness needed in various situations. If such rules exist they can be applied to the new application. If none exist, the appropriate members of management must define the rules to be applied. Care must of course be taken to ensure that realistic rules are defined; integrity requirements which are too rigid can increase system cost unnecessarily; those which are too tolerant can decrease system accuracy unacceptably.

System-availability requirements must also be defined, and this is another area in which care is needed. Data-processing professionals understand that 100 percent uptime is exorbitantly expensive and ultimately unattainable. (No matter how much redundancy there is in a system, there is always some possibility of multiple failures which cause a system outage.) The cost of improved availability tends to grow geometrically, at least in a centralized system, as availability increases from 99 toward 100 percent. The key to keeping costs in line while providing the uptime needed is to define system-availability requirements precisely. It is important, for example, to define the periods during which high availability is needed; usually these are the peak activity periods and occur during the daytime business hours. It is also essential to realistically define the length of system outage which could be tolerated, even during a high-availability period. As an example, in airline-reservation systems a second computer is typically used to back up the reservation-processing computer. If a failure occurs, causing switchover to the backup system, there may be up to 30 seconds (depending upon the system and the switchover method used) during which the reservation system is inoperable. Even in this high-volume, on-line application, such a short outage can be tolerated.

It is useful to explore alternatives when discussing availability with the users and their management. For example, if a person will use the same terminal to access a database and also to create memos and then transmit them via electronic mail, it might be useful to suggest that if a problem develops in database access, memo creation and transmission can continue uninterrupted. Or perhaps both database access and mail facilities might be affected, but the creation of memos could continue. If the users can cope with selective outages, the next step is to determine how long each failure could last before operations are seriously affected. If these selective outages appear to cause operational difficulties for the users, it would then be appropriate to try to arrive at the monetary value of a backup facility to avoid the disruption caused by either or both types of outage.

Assigning monetary values is exceedingly important, since high up-time can be expensive—and always involves *some* extra cost. If a value can be placed on each possible backup or fallback option, then the cost of those options may be justified. Of course, if the cost of a specific backup option is high but its value to the organization is low, that leads to a straightforward decision during system design. The quantification of requirements, whenever possible, is good practice, since system-design decisions can be made much more easily when the value of a specific function or feature is defined quantitatively.

SECURITY REQUIREMENTS

Security requirements may originate in legal constraints, in contractual commitments, or in the need to protect data and/or computer facilities which are of value to the organization. During data collection the emphasis must be on defining legal, contractual, and data-protection needs. The protection of computer equipment, both physically and from inappropriate use, must be defined during system design (Chapter 13), because the need for this type of protection is not defined in terms of user requirements.

Legal requirements for security protection often affect governmental entities. The U.S. government practice of classifying data into categories such as confidential, secret, and top secret—with a defined level of protection required for each category—reflects legislation in this area. New laws, especially in countries other than the United States, are being enacted to prevent the flow of certain types of data across national borders; this is often referred to as the control of transborder data flow. Information on legally mandated security protection can best be obtained from the organization's legal staff or from the appropriate members of management.

Many organizations overlook the fact that security-protection requirements may be defined in contracts. As an example, a manufacturer might obtain product-related data from its suppliers or subcontractors and be contractually committed to use those data only internally and without disclosure to a third party. There may, in some cases, be a commitment that the supplier's or subcontractor's data will be made available to only those employees who have a specific need (the "need to know"). In the latter case, protection for the data may be more stringent than for many of the organization's own data elements. Contractual requirements which affect the new application can be determined through interviews with management and/or the legal staff.

Other sensitive information which is of value to the organization and of potential value to other organizations or people must be identified during data collection, so that appropriate protection measures can be defined. There has been a great deal of publicity concerning computer fraud involving funds-transfer systems, so there is currently a high degree of awareness that financial data—especially when being transmitted between different institutions—require tight protection against misuse or disclosure. There is perhaps less awareness that many other types of data may be equally sensitive. For example, if a business sends out advance information on price changes, in a highly competitive industry this information might be of considerable value to competing organizations. It may be possible to place a value on the ability to change prices without the knowledge of competitors until the change is effective, and this value will be very useful in justifying security-protection methods to avoid disclosure. Organizations involved in research, particularly in high-technology areas such as genetic engineering and microelectronics, often maintain extremely valuable information about techniques, processes, or products which are not yet protected by formal methods such as patents. These data may be of great value to competing organizations, and it may be possible to assign quite specific values based on the amount of time and effort expended in the related discoveries.

PRIVACY-PROTECTION REQUIREMENTS

Privacy-protection requirements may be legally dictated or may arise from public sensitivity to the misuse of personal data. Any data describing personal characteristics, which are stored or processed by a computer-aided system, may generate the requirement for privacy protection. Certain types of personal data, especially those which describe an individual's financial status or liabilities and those which relate to an individual's medical status or history, are considered to be particularly sensitive.

Legal requirements to protect personal data vary. In the United States, within the private sector (nongovernmental organizations) legislation at present covers the protection of credit status and history data. This legislation is concerned with the accuracy of credit data and allows a person to correct or register a protest concerning credit data which he or she believes to be erroneous. This legislation was intended to correct earlier abuses: a person's computerized credit records were often unavailable for that person's review and might contain innuendos or false statements negatively affecting the person's ability to obtain credit.

The U.S. Congress has passed more wide-ranging legislation concerning the protection of personal data in computer-aided systems which are operated by government organizations. This legislation provides that an individual must be notified of which personal data elements are maintained in computer-based systems and the purpose(s) for which those data elements are used. It also provides for notification of the individual if additional data elements are acquired and/or if the data are used for new purposes. There have been several attempts to pass legislation which would apply the same protection rules to data stored by private-sector organizations, but action has not yet been completed. It is likely, however, that legislation of this type will eventually be enacted.

In addition, there is growing public awareness of the amount of personal data maintained in computer-based systems, with growing concern about the possible misuse of those data. In general, therefore, personal data—especially financial and medical data—require protection whether or not there are specific legislative requirements.

Legal requirements for privacy protection can be determined by consultation with the legal staff and/or management. If there are no legal requirements, the types of data elements which require privacy protection are usually self-evident. Additional requirements, perhaps arising from specific sensitivities among the organization's work force or customers, must be determined by consultation with management and/or the system users during data collection.

AUDITABILITY REQUIREMENTS

Auditability requirements must be carefully defined in cooperation with the internal auditing staff or the external auditors. As in the case of security and privacy, there may be legal requirements that specific records be kept for auditing purposes. In addition, there are accepted accounting practices which provide for auditability; too often, those practices have been ignored in the implementation of computer-aided applications. Auditability is related to integrity, as one aspect of auditing is to ensure that the records of the organization have an adequate degree of accuracy. Auditability is also related to security, since one purpose of an audit trail is to provide a record of which transactions were initiated, who initiated them, and what the results were in each case. On-line applications too frequently leave no audit trails, making it difficult to determine the accuracy of the processing as well as making it possible for transactions to be improperly initiated for financial gain.

There are different philosophies concerning the auditing of computer-based systems. In some cases auditors have avoided becoming involved in system definition and design to preserve their independence and objectivity. When this approach is taken, the auditors are usually unable to determine if the system operates correctly or if intruders (or authorized personnel with dishonorable motives) could cause the system to operate to their advantage. Auditing philosophy today is shifting to the belief that the auditors must be involved in the system from its inception and be skilled enough in data processing to be able to review the design and implementation to determine if auditing requirements are met. This approach is preferable, since auditability is an important aspect of computer systems which are involved in an organization's main-line operations. Also, few system designers are technically qualified to design procedures which will ensure that auditing requirements are met. Considering the auditors, internal and/or external, to be an integral part of the design team is the best way to ensure that those needs will be taken into account.

DOCUMENTATION

Documentation of the requirements for integrity, security, privacy, and auditability is not easy to formalize, because there are so many different requirements in different applications. However, the requirements for integrity, security, and privacy protection are related to the data elements—particularly financial, quantitative, and personal data elements—stored and/or processed by the system. Documentation can therefore be organized in terms of those data elements, as shown in the example in Figure 5-5 earlier in this chapter. Note that the example shows one data element, the *process sequence*, which requires high integrity protection and also security protection. The sequences of processes by which items are manufactured are often proprietary and of considerable value. They must also be exactly accurate, or the finished items will not be satisfactory. All other items are noted as requiring normal protection and accuracy. Auditability requirements can generally be defined in narrative form.

EXPECTED CHANGE

Every effort must be made during data collection to determine which aspects of the system being defined are likely to change. Although system design must aim at flexibility for change in all aspects of the application (see Chapter 13), it is useful to know which types of change are most probable.

Some possibilities for change are evident. Prices and wage rates change continually. The rate of payroll deductions for U.S. social security payments and the maximum amount of salary to which deductions apply change almost every year. The majority of the monetary figures in an application, in fact, can be expected to change—if not regularly, at least occasionally. It is simple to design so that these changes can be made easily.

Other types of change are less evident and may also be more difficult to handle. Examples are modifications in the way in which business is done or work is handled. With careful analysis it is often possible to evaluate changes and discover that the basic differences are not as great as they seem. As an example, a purchase-order system might be changed so that instead of issuing purchase-order documents to suppliers, the appropriate purchase-order information is transmitted electronically to each supplier's computer system. While this is a significant change in the way purchases are handled, the same type of information is required in each case: the identity and quantity of each item to be purchased, the identity of the purchaser and of the supplier, the required shipment date(s), and an authorization of the amount to be paid when the items are delivered. The method of output changes from hard copy to transmission via data communications (a proof copy may be printed locally for use in the purchasing department), and it is likely that the format of the purchase order will also change. In fact, a different format might be required for each supplier, to match the formats of that organization's information system. If each function of the purchase-order application is designed modularly, this type of change can be accommodated. The module which generates the purchase-order data will remain the same, while the formatting and output-preparation modules will change.

During discussions with the prospective users of the system, it is helpful to ask about past changes. Past experience often indicates which types of change are typical in the application being studied. Information is needed about not only the types of change but also the frequency of change and the interval which occurs between the decision to change and the implementation of the change. The latter is important because computer-based information systems must be carefully designed if a change is to be implemented very rapidly after the decision to change is made. There are many cases in which this capability is vital. Business is often very competitive, and the decision to change prices or the terms of sales must be implemented quickly. Special emphasis must therefore be placed on changes which must be made rapidly.

System users may not always be the best source of information about possible future change, and they are certainly not the only source. Management, the legal staff, and the auditors ought to be consulted concerning the types of change most likely to occur. Reorganization of the structure of the enterprise, possible mergers or acquisitions, expected legislation affecting the organization or the application under study, and changes in rules such as those defined by the Securities and Exchange Commission (SEC) in the United States for the computation and disclosure of company financial information are examples of changes which may be predicted. In other cases, there is only a possibility that change will occur.

Forecasting the future is a very tricky business, and no one has a magic formula for doing so with complete success. However, the more information about expected change that can be gathered, the more likely that the design of the new system will be amenable to modification in appropriate ways. It is also fruitful to analyze various parts of the application, related work flow, and related organizational structures of the enterprise and imagine what types of change *could* occur to affect each aspect. Some of these speculations may be far-fetched, but others may open up possibilities which would otherwise be ignored. For example, suppose that the organization decided to move aggressively toward a very flexible work-force arrangement, with people working dramatically different hours from their present schedules and/or perhaps with many people working in their homes. What effect would this have on the application? Could this type of change be accommodated, and if so, how?

Partial documentation of the expected requirements for change can be accomplished with the same forms used for other aspects of data collection. The documentation of output requirements (Figure 5-1), geographical locations (Figure 5-4), and personal-data sets (Figure 5-6) includes in each case an entry for "other information," in which expected changes can be noted. The form documenting data elements in Figure 5-5 also provides a column in which Y or N can be noted to indicate if modifications are expected. The documentation of processing flow (Figure 5-3) must also be annotated to show expected, probable, or possible change. In some cases these annotations may be adequate; in others it will be necessary to append a narrative description of the type of change expected. If this is done, a note number can be inserted in the change column or "other information" to cross-reference that item to the associated narrative.

MANAGERIAL CONTROL

The final task in data collection is to determine the managerial philosophy and style of the organization, or of that segment of the organization related to the new system. Management philosophy may be well known and in fact may be formally documented in the procedures and policies of the organization. In other cases there may be no documentation describing the management philosophy, and it may be difficult to discern exactly what the management style is. (This is especially true in many data-processing organizations, which can become quite "cloistered" and isolated from the operational organizations of the enterprise.)

Management style has typically been ignored in analyzing and designing computer-based systems, often with unfortunate results. When computer systems were almost always centralized, the control of data-processing was also centralized, regardless of how the organization as a whole was managed and controlled. Many organizations still operate centralized information systems, and many experience a continual tug-of-war between the centralized control of computer resources and distributed control of the line organizations. There are, of course, organizations which are heavily centralized in their management philosophies; this is especially true of small, privately owned businesses. In these cases a centralized data-processing structure exactly matches the organization's management philosophy.

Larger organizations, whether in business, government, health care, education, etc., tend to have either formal or informal distribution of a considerable degree of management responsibility and authority. This is especially true in the United States, where a major reevaluation of management principles is in process. Assigning the maximum amount of responsibility and authority as far down in the organization as practical is currently believed to be the best way to achieve flexibility and improve productivity.

Distributed, or localized, management control tends to encourage the formation of distributed or decentralized information-processing systems. Local managerial control generally means that each local manager requires somewhat different information for decision making and may use different methods to maximize efficiency. Forcing all organizations into a common pattern, as is typical with centralized data-processing systems, will be very much at odds with this type of organizational structure and philosophy.

On the other hand, even in organizations with a great deal of local management authority, there is usually a requirement for some types of review and decision making at a higher management level. As a minimum requirement, there is a need for financial data at the corporate or headquarters level to reflect the status of each lower-level component and to allow data consolidation for reporting purposes, tax preparation, stockholder reports, and so on. This tends to lead to links from the local information-system components to some central-processing location, where those functions needed by upper-level management and the financial organizations at headquarters are performed.

These remarks are, of course, generalizations from patterns observed in many organizations. It is important to determine exactly what pattern of management control is used in the specific organization and how this affects the application being studied. In some cases there may be no impact, but in others knowledge of the management philosophy will have a definite effect on the design choices of system structure (Chapter 10) and database structure (Chapter 11).

A list of sample questions follows which can be asked during interviews with management as part of the data-collection process. The answers to these questions will generally lead to an understanding of the management structure and how that structure will affect the system design of the new application. The answers supplied will be adequate documentation of the managerial-control aspects of the new system.

1. What level of management makes purchasing and investment decisions, and for what amounts of money?

   • The lower the level at which decisions are made and the higher the purchase or investment amounts allowed at that level, the greater the distribution of management authority.

2. In a for-profit organization, where does the responsibility for profit and loss lie?

   • Profit-and-loss responsibility at divisional or departmental level, or even at an operating-unit level, indicates a considerable degree of distribution of management authority.

3. At what level are organizational policies defined, and for which types of policies?

   • The lower the level at which significant operating policies are established, the greater the distribution of management authority.

SUMMARY

The process of data collection, as described in Chapter 4 and this chapter, is one of the keys to the successful design and implementation of a complex system. It is not unusual to spend a month interviewing prospective system users; for a large, complex system which involves changes in the way work is done, a data-collection phase lasting 3 months is not excessive. This does not mean that data collection goes on 8 hours a day, 5 days a week, for a month or for 2 or 3 months. There are typically interruptions, and the users can probably spend only a limited amount of time each day or week interacting with the system-design staff. These are therefore elapsed times, not measurements of continuous effort. There are many systems in which much longer than 6 months was spent in the data-collection process, and this extra effort has typically been repaid in easier completion of the remainder of system analysis and design.

It is extremely difficult for many system analysts to avoid becoming impatient with the data-collection process and also with the tedious job of fully *and clearly* documenting the data accumulated during that process. It has been estimated that only 3 percent of the total system-design and implementation time and effort are spent on the analysis of system requirements. In a simple system, for which the requirements may be well-defined and self-evident, this may be adequate; in a complex system it is clearly inadequate. If the data-collection process is not done well, the system requirements will be poorly defined and the resulting system may be unsatisfactory.

There is growing agreement within the information-system industry that the *front-end* functions of data collection, requirements definition, and system analysis are the most important in the entire sequence leading to implementation. This is an extension of the earlier understanding that system design is more important than coding. In computer-based systems in the 1960s, little time was spent on design but a great deal of time was spent on coding. This often led to repeated recoding as the design was refined during implementation. Emphasis then shifted to completing a more detailed design before any coding began. Today the emphasis is shifting still farther toward the beginning of the process, to focus on requirements definition and system analysis. This shift in focus is especially important as the new types of applications described in Chapter 1 are implemented. Those applications involve a great deal more process analysis and interaction with the people who will use the system than was typical in computer-based applications 10 years ago.

The message, therefore, is that adequate time and effort must be invested in the data-collection process, to ensure that the needed data are obtained and formally and clearly documented. These data form the base upon which all the later steps in the analysis, design, and implementation process will be built.

6

REFINING
APPLICATION
OBJECTIVES

When data collection is complete, it is time to refine the objectives to be met in the implementation of the new application. The implementation-oriented objectives must be defined, and in some cases the strategic objectives may now be modified. The understanding of the requirements ought to be much improved as a result of the interaction among users, system designers, and management during data collection.

Implementation-oriented objectives are stated in terms which are relevant to the data-processing professionals who will implement the necessary programs. These objectives are also directly related to the users' view of the system and what the system must supply to assist them in carrying out their duties. This dual purpose—to be relevant to both users and implementers—must be kept constantly in mind. The definition of objectives forms the "contract" between the users and the designers and implementers, and it must therefore be understandable and meaningful to both groups. An example set of objectives is provided as part of the case study in Chapter 16.

DOCUMENTING THE OBJECTIVES

Most or all of the information necessary to define these objectives ought to be available as the result of the data-collection process. What remains to be done is to organize that information into a coherent document for review by the users and their management. The following items make up the implementation-oriented objectives:

• System description

• Functions and functional flow

- Response-speed objectives

- Availability requirements

- Integrity, security, privacy, and auditability requirements

- Flexibility and adaptability objectives

- Cost limitations

- Schedule

Availability requirements were grouped with data-integrity requirements during the discussion in Chapter 5 of data collection. However, availability is so important that it is best to define it separately in the statement of objectives.

The following paragraphs describe each of the implementation-oriented objectives and how each can be documented.

SYSTEM DESCRIPTION

A system description, in narrative form, is the first item needed. This narrative must provide an overview of what the system will do, how it will operate, and the benefits expected to be realized from its implementation.

As an example, a new system in a hospital might be designed to track patient status. It would include methods for entering patient data on admission as well as provisions for entering all required medication and treatment information. It would also provide for messages to the pharmacy about medication needed and to the laboratory for specific tests required for each patient. It would allow for entering patient-release data, and might at that time require the transmission of data to another system (or application) for preparation of the patient's bill. It might also offer a variety of inquiry facilities—to determine patient status, empty beds, scheduled tests, and other similar information.

The system description for this hypothetical hospital application would include all the above information and would also indicate generally how each type of data would be entered and how output information would be used. For example, personnel in the admissions office might use a terminal to determine available beds, then enter admission data for each new patient at the time of arrival at the hospital. Other data might be entered by nurses, nurses' aides, doctors, or lab technicians. This description need not be in detail and need not describe the system work flow; that will be defined under Functions (which follows). The system description must provide an introduction to the statement of objectives and must be in terms easily understandable to non-data-processing personnel.

The final entry in this section of the objectives must describe the expected benefits. As discussed in Chapter 3, benefits may include lower costs, greater efficiency, improved productivity, increased strength in competitive markets, and any other benefit(s) appropriate to the system and the organization. Quantitative measures of expected benefits are generally included in the strategic objectives, but they may alternatively be defined here if it is more convenient to do so.

FUNCTIONS

Functions to be provided must next be defined in overview form and from the user's point of view. It is unnecessary to define system-support functions which are not apparent to the user; for example, an operating system with specific characteristics will be required, but this is (or ought to be) invisible to the user of an interactive system. Information about secondary or support functions (operating software, utility routines, and so on) which is obtained as a by-product of data collection ought, however, to be documented for later use, as it will be helpful in estimating the time and effort required for system implementation.

The work-flow analysis performed in Chapter 4 will provide some of the information needed to describe functions. The other source of functional requirements is the description of processing needs obtained during data collection (see Chapter 5, Figure 5-3).

The functional objectives are best described using an annotated flow-chart in a format similar to that shown in Figure 5-3. In addition to the indication of the functions and their flow, the users of each set of functions must be defined. It is also necessary to specify the usage modes (see Chapter 5, Figure 5-2). Finally, the major types of output must be shown, with an indication of where the output will fit into the flow and who will use each type of output. (See Chapter 5, Figure 5-1, for the original definition of output requirements.)

RESPONSE SPEED

Response-speed requirements must also be included in the statement of objectives. These requirements can be obtained from the data collected in the process described in Chapter 5 and can be transferred from the documentation prepared at that time (see Figure 5-1). A system may include some output reports which do not have specific response-speed, turnaround-speed, or due-date-and-time requirements. If this is the case, it must be noted, in order to avoid the inference that the requirement has been unintentionally omitted.

For each requirement, a proposed method of measurement is usually needed. In some cases, such as long-turnaround batch processes or due-date reports, either measurement is not worthwhile or the methods are self-evident. (If, for example, a payroll is due each Friday at noon, any failure to meet that schedule will be readily apparent.)

Care must be taken to define measurement methods which are objective and which also use representative samples. Especially in the case of fast-response needs (2 seconds or less), it is best to provide measurement tools built into the system software and/or hardware. Since those measurement tools will unavoidably add to system overhead, it must be possible to enable them only when needed. For less stringent requirements, which usually include more latitude, manual timing studies may be adequate and less costly than automated methods. Also, methods such as timing with a stop watch are more accurate in measuring several-second response than shorter response times.

Response-speed measurements must be made during the final tests of the system prior to production operation. These measurements, however, may not be representative of the fully operational system. Additional measurements must be made, both during the phase-in of the system to operational use and after full production is achieved. Monitoring the speed of response is a continuing process, since many factors can degrade response and affect the system's ability to properly serve its users.

AVAILABILITY

Availability requirements form another part of the application objectives. The information necessary to define these requirements is obtained during the data-collection process. (During that process, availability was grouped with integrity because of their close relationship.) Availability objectives must define the following for each major system function:

- How long an outage can be tolerated by an individual user, by a class of related users, and/or by all users
- How frequently outages can be tolerated, for the same groups as above
- What backup or fallback methods will be used to handle each type of outage
- How availability will be measured when the system is operational

Defining availability in the users' terms, i.e., frequency and length of outages, instead of in technical terms, e.g., 99.2 percent availability, and for each major function separately, is essential. A given availability percentage can be achieved in more than one way. For example, 99 percent availability in a system which operates on-line 12 hours a day, 5 days a week, can be achieved if there is one outage a month lasting 2 hours and 24 minutes. It can also be achieved with one outage a week, averaging 36 minutes. In some cases either situation is acceptable; in other cases outages as frequent as one a week will not be tolerable.

As in the case of response speed, it is appropriate to define multiple points at which availability will be measured. Initial measurements during phase-in of the system to operation will be useful but are no substitute for continuing measurement and evaluation during production use.

INTEGRITY, SECURITY, PRIVACY, AND AUDITABILITY

Integrity, security, privacy, and auditability requirements must be included in the implementation-oriented objectives. The data-collection process described in Chapter 5 ought to have defined those requirements completely, and that information can be transferred from the earlier documentation to the objectives statement. Detailed information, such as the exact level of integrity needed for specific data elements, is not appropriate in the statement of objectives. Only major items, such as functions or data which require special security or privacy protection, plus a narrative describing auditability needs, are required. However, it is a good idea to attach the more detailed documentation prepared during data collection (Figures 5-2 through 5-6) to the statement of objectives as appendixes which can be reviewed by those interested.

FLEXIBILITY

Flexibility requirements will also have been defined during the data-collection process. Information about expected changes and the types of modifications most likely to occur must be included in the statement of objectives. To reiterate, it is impossible to define all changes which will occur; however, the most complete practical statement of likely changes will assist in designing the system to readily accept modifications.

The required speed of implementation when change occurs must also be included. While this may be difficult to define, it is essential. Realistic trade-offs must be made between user management wishes, which are generally in favor of rapid change when modifications are necessary, and the time required for implementation. A close working relationship between users and system designers, which ought to have evolved during data collection, will aid in defining objectives for the speed of change which can be accepted by both groups.

COST

Cost limitations must be formally specified in the objectives. An estimate of the cost of design, implementation, and operation is usually made during ROI calculations (see Chapter 2). There are, however, some applications whose implementation is mandatory and which therefore bypass the steps that include computing ROI. After data collection, a more accurate cost estimate can typically be made, and this must be included in the objectives. In this approach, the estimated cost is proposed by the design staff, to be either accepted or rejected by user management and/or IRM management. Another approach is to use budgetary guidelines which are established in some IRM organizations. For example, the budget for new application development might allow three people to be assigned to this application, resulting in a known cost and an estimated time of completion.

SCHEDULE

Schedule goals are also important to indicate when (parts of) the application will be usable. The required schedule may be stated by management and if so, must be reflected in the objectives exactly as stated. If no schedule has been specified, a proposal must be made by the system designers.

Schedule and cost can be traded off in many cases; higher expenditure *may* produce a shorter schedule (although this is by no means automatically the case). Limited funds may of necessity dictate a longer, phased implementation schedule. If there are trade-off possibilities, they can be presented in the review of objectives, so that user management and IRM management can decide whether prompt implementation or lower cost is more important.

REVIEWING THE OBJECTIVES

The next step is to formally review the objectives with the users and their management. This is best accomplished through a combination of presentation and written review. If the users have been involved throughout data collection and system definition, the review process may be quite simple, with almost automatic agreement. In other cases, the objectives may be revised and refined during the review.

It may be helpful to schedule a presentation for the users and their management and to describe the system objectives (both strategic and implementation-oriented) during that review. Visual material covering the system functions, flow, output, response, and availability goals is appropriate. An informal atmosphere, with discussion and questions, must be encouraged. However, this process is also part of selling the users on the new system; therefore proper preparation and well-done visuals will be worth the effort. After the presentation and discussion,

---

XYZ CORPORATION
Anytown, U.S.A.
SYSTEM OBJECTIVES AGREEMENT

September 1, 1982

The undersigned have agreed on the statement of information-system objectives fully defined in the enclosed document titled "Manufacturing Process Analysis Application Objectives." That statement of objectives accurately reflects the business situation and the requirements of the future users of the new (revised) information system. The methods for measuring the adequacy, accuracy, and effectiveness of the system when implemented are also defined in the enclosed statement of objectives.

The Information Resources Management Department agrees to implement the necessary computer-based procedures to realize these objectives, with a target installation date of March 31, 1983 for cutover to production use. The estimated charges for this implementation will be $50,000. To aid in the implementation process, the Manufacturing Department agrees to provide the equivalent of two full-time manufacturing analysts as consultants to the IRM Department.

Although the undersigned agree that this statement of objectives reflects their best current understanding of the requirements, these must be considered as subject to change if external conditions change and/or upper management makes certain executive decisions. In such a situation, this agreement will be renegotiated as soon as possible to reflect the new conditions.

_____   _____   _____

Director, Manufacturing   Director, IRM   Project Leader,
                                                      System-design team

---

Figure 6-1 Example agreement on system objectives.

representative users and their management can be given the written statement of objectives for review. Any disagreements must then be resolved and reflected in a revised document.

Upon agreement, a formal "contract signing" is appropriate. User management and the management of the system-design staff and/or the IRM department, as well as the lead system designer(s) for the project, are the parties to this agreement. The agreed-upon statement of objectives can be accompanied by a cover sheet similar to the example shown in Figure 6-1.

The statement in the example concerning how change will be handled is important and must not be omitted. A mutual understanding of the possibility of change without notice and its potential impact on the system-implementation process is essential to a successful information system.

DATA ANALYSIS

7

DETECTING PATTERNS

The next step in system analysis and design is to analyze the data which have been collected. Because this analysis provides essential information as input to system design (Section 5), it is important not to bypass the data-analysis step—as, unfortunately, is often done. In the study of a large information system, the data collected to describe the system requirements, users, and other factors are always voluminous. If these data are not thoroughly analyzed, important aspects of the requirements and/or of the environment within which the information system will function may be overlooked or—even worse—misunderstood. Care in determining exactly how the data describe reality will be well worthwhile.

The term "pattern," as used in this book, describes a relationship or configuration which exists in the real world and which is relevant to the information system being analyzed. For example, the relationship of certain workers to the department within which they work is one type of pattern. The relationship(s) of those same workers to the type(s) of work they do forms one or more additional patterns (depending on whether or not all workers carry out the same tasks). Because the information system must fit into the users' environment, relationships which exist within that environment are germane to the design of the system. In addition, many relationships can affect how the system is designed.

It is one of the postulates of this book that the use of patterns which exist at present or which will exist in the future is the best base for a good system design; that is, a design which accurately reflects system requirements. This chapter shows how the data collected as described in Chapter 5 can be analyzed to discover various types of patterns, while Chapter 8 tells how those patterns relate to different aspects of system design. Section 5 then builds on this foundation to describe the process of strategic-level system design.

The patterns which are helpful during system design and which can potentially be found in the collected data follow.

- User-group relationships
- Data-access needs
- Geographical groupings
- Integrity-level requirement patterns
- Security- and/or privacy-requirement patterns
- Management-requirement patterns
- Patterns related to expected or potential system change

USER-GROUP RELATIONSHIPS

There are typically multiple ways to group the people who will use the new (or modified) information system. In the process of identifying system users described in Chapter 4, the emphasis is on categorizing users in terms of shared functional needs. In an airline-reservation system, for example, personnel working in a telephone-based reservation center have a common set of functional requirements. Personnel working for the same airline at an airport or city ticket counter may share some of the same requirements but may have additional needs unique to their duties and the working environment. In this case the latter group of users is defined as forming a new pattern.

Another way to categorize users is by grouping those with common requirements for access to data. The resulting patterns may or may not exactly agree with the patterns determined by functional needs. In the airline-reservation example used earlier, both groups of users (those in the reservation center and those at ticket counters) would require access to the same set of flight-bookings data. The single data-access pattern would therefore overlap the two user-group patterns. In a different example, all agents of an insurance company might share common functional needs but each would have a requirement for access to the records for his or her own clients. Note that this pattern, like some others, may not prove to be pertinent or useful later in the system design. However, during the analysis of data patterns it is important not to prejudge which patterns will be important and which will prove uninteresting. It is better to analyze too thoroughly (if that is possible) than not thoroughly enough. The next section discusses data-access-need patterns in more detail.

Still another way to group users is by their geographical locations. The term "geographical" must not be taken too literally, in the sense of widely dispersed locations. Users may be grouped by the office in which they work, by the floor on which the office is located, or by the building within the city, as well as in more geographically separated patterns. A later section discussed the analysis of geographical patterns.

These are not the only possible ways to group users, but they are the ways found in most systems. When a specific set of data is being analyzed, other user-group patterns may be apparent. In most cases, the analysis of user-group patterns will produce an overlapping set of groups, as shown in the example in Figure 7-1 (which is a simplified excerpt from a typical documentation form). An individual user may participate in one, two, or more groups.

APPLICATION: RESERVATION PROCESSING

1. Reservation clerks
   1.1 Telephone-reservation clerks
       1.1.1 Omaha
   1.2 City reservation offices
       1.2.1 New York City, Sixth Avenue
       1.2.2 New York City, Olive Street
       1.2.3 Boston
       1.2.4 Washington, D.C.
       (etc.)
   1.3 Airports
       1.3.1 JFK
       1.3.2 La Guardia
       1.3.3 Logan Field
       1.3.4 Dulles
       (etc.)

   Data-access requirements identical

2. Supervisors
   2.1 Telephone-reservation center
       2.1.1 Omaha
           (*Note:* Located on floor above reservation clerks)
   2.2 City reservation offices
       2.2.1 New York City, Sixth Avenue
       2.2.2 Washington, D.C.
       (etc.)

   Data access:
   1. Identical with reservation clerks
   2. System statistics showing how system is operating
   3. Statistics for each clerk's transaction rate, error rate, etc.

Figure 7-1 User-group patterns.

DATA-ACCESS NEEDS

Patterns in data access form an important element in system design. There are a variety of ways in which access patterns can be defined. One is in terms of which user groups need access to specific "pools" of data, or sets of data elements—which results in patterns very closely related to the user-group patterns discussed above. Another way to group data accesses is by the time period when specific types of access will occur. For example, a group of terminal users may access customer files interactively during the business day, and batch-update programs may access the same files during the night. This would result in two patterns, one to reflect the users' daytime interactive requirements and the other to reflect the batch programs' accesses at night. Some accesses may occur weekly, monthly, quarterly, and/or annually. Each of these will define a data-access pattern consisting of specific data elements.

As the earlier example indicates, another type of pattern concerns the way in which specific data elements are accessed; interactive and batch-mode access are the two types to be considered. Each type of data access must be fitted into one of these categories. Finally, data-access needs can be grouped by those which occur as the result of predefined production processes and those which are ad hoc. The latter patterns are of course very difficult to predefine, but an attempt must be made, based on the data collected earlier.

The result of this analysis of data-access patterns will be multiple overlapping groups of data elements. In each case, a specific group (or pattern) of elements will be related to (1) access by a user group or program, (2) access during a specific time period, (3) access either interactively or in batch mode, and (4) access because of a predefined production process or in ad hoc mode. Figure 7-2 provides a simple example of how data-access patterns might be documented to show these interrelationships.

GEOGRAPHICAL GROUPINGS

The system design will of course be affected by where users and system equipment are located. The existing situation must be analyzed for patterns, and possible future changes in locations must also be probed.

The locations of system users form one set of patterns. Since users are generally "fixed," in that they occupy specific offices, factories, or other facilities, these are predefined endpoints in the information system to be designed. These locations must be documented as part of data collection and analysis. User locations must be defined in terms of the office, factory section, etc., rather than simply in terms of the city. In many cases the more specific information will be needed when network design, in-

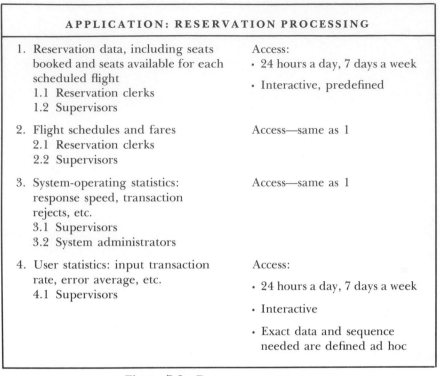

APPLICATION: RESERVATION PROCESSING

1. Reservation data, including seats booked and seats available for each scheduled flight
   1.1 Reservation clerks
   1.2 Supervisors

   Access:
   • 24 hours a day, 7 days a week
   • Interactive, predefined

2. Flight schedules and fares
   2.1 Reservation clerks
   2.2 Supervisors

   Access—same as 1

3. System-operating statistics: response speed, transaction rejects, etc.
   3.1 Supervisors
   3.2 System administrators

   Access—same as 1

4. User statistics: input transaction rate, error average, etc.
   4.1 Supervisors

   Access:
   • 24 hours a day, 7 days a week
   • Interactive
   • Exact data and sequence needed are defined ad hoc

Figure 7-2 Data-access patterns.

cluding the design of local-area networks, begins (see Chapter 12). Of course, some users may not have fixed locations, at least in terms of their relationship to the information system. Examples are traveling salespeople, who may use portable terminals or telephones to input data to the system and to receive output. Other examples might be professionals who, when traveling, access the electronic-mail functions of the system to receive or send messages. In most cases, all traveling users can be considered to form a single group for system-design purposes, although if the requirements vary considerably, it may be appropriate to define multiple groups.

Geographical groupings of computer components which are already installed must be identified as well. Facilities and/or locations which are suitable for the installation of computer equipment of various types must be analyzed during this process. All this information will typically have been gathered during data collection and must now be documented in terms of geographical patterns.

Finally, existing private data- and/or voice-communications facilities must be analyzed. Many organizations have private networks installed,

Figure 7-3 Geographical patterns.

often entirely for voice-communication use. Information about such a network and its geographical relationship to the locations of the users of the new system will be useful during system design.

The result of this analysis of geographical patterns will be a map or maps showing major clusters of users and existing or potential equipment sites, with existing or planned communications facilities. Supporting documentation is needed to identify the items shown on the map(s). Figures 7-3 and 7-4 are examples of how these patterns can be documented.

INTEGRITY-LEVEL
REQUIREMENT PATTERNS

All the information necessary to define patterns of integrity protection ought to be available when the data-collection process discussed in Chapter 5 is complete. The patterns of interest are those which separate data elements which require a high level of integrity protection from those which do not. In some systems there may be more than two groups, further categorizing various possible levels of protection. Typically, however, data elements require either the highest feasible level of integrity protection or little or none. For example, new applications often include text or document processing and management, which may involve a very large volume of text but which may not require the type of integrity protection typically applied to databases. Textual material is usually intended to be read at some time(s), and the reader can cope with inaccuracies such as misspelling, altered characters, etc., very nicely in most cases.

The necessary data to define these patterns can be found in the documentation completed in Chapter 5, as shown in the example in Figure 5-5 of data-element definitions.

SECURITY- AND/OR PRIVACY-REQUIREMENT PATTERNS

The data-element documentation shown in Figure 5-5 is also the source of patterns in security and/or privacy protection. However, more sepa-

---

APPLICATION: RESERVATION PROCESSING

1. Omaha
   1.1 144 State Street
       1.1.1 Telephone-reservation center (4th floor)
             1.1.1.1 Reservation clerks
       1.1.2 Supervisory center (5th floor)
             1.1.1.2 Supervisors
   1.2 418 Central Avenue
       1.2.1 Computer center (basement)
             1.2.1.1 System administrators
             1.2.1.2 Operations staff
       1.2.2 Development center (floors 2 and 3)
             1.2.2.1 System analysts
             1.2.2.2 Programmers

2. New York City area
   2.1 1811 Sixth Avenue
       2.1.1 Reservation office
             2.1.1.1 Reservation clerks
             2.1.1.2 Supervisors
   2.2 446 Olive Street
       2.2.1 Reservation office
             2.2.1.1 Reservation clerks
   2.3 JFK Airport—Terminal #14
       2.3.1 Reservation clerks
       (etc.)

3. Boston area
   (etc.)

Figure 7-4 Geographical locations.

APPLICATION: FACTORY-FLOOR MANAGEMENT

1. Highly secure data
   1.1 Process sequence

2. Secure data
   2.1 Bill-of-materials relationships
   2.2 Supplier identification and contract terms

3. Private data
   3.1 Employee work rates
   3.2 Employee pay rates

All other data elements in this application require no special protection for security or privacy.

Figure 7-5 Protection patterns for security and privacy.

rate patterns, or groups, of data elements will typically result when security and privacy are analyzed than when integrity is involved. In some systems, there may be few requirements for security protection and no privacy-protection needs. In others, there may be several classes of security protection—as in U.S. government data-sensitivity classification schemes—each of which forms a pattern for data analysis and at least one class of privacy protection. It is also possible to have multiple privacy-protection groups. For example, personal data may require protection against access by unauthorized persons or programs and may require a different level of protection against unauthorized change. An example of how security-protection (or privacy-protection) patterns might be documented is provided in Figure 7-5.

MANAGEMENT-REQUIREMENT PATTERNS

Many information systems include patterns based on management requirements and/or on management's need for control. For example, in some systems certain data elements are required for executive-level reports. Those data elements form a pattern based on executive interest. Other data elements will be required for use in reports and/or inquiries by division-level management, and therefore form one or more additional patterns. The executive-level data elements and division-level data elements may form overlapping patterns; this is true in most organizations.

Management's data needs are a special facet of the more general data-access patterns discussed earlier. They are categorized separately because of their importance to the organization and also because they may include more frequent change, a higher degree of ad hoc access, and a greater need for data protection than the data-access needs of other users.

In addition to patterns formed by data-access requirements, there may be patterns formed by grouping functions which are of interest to various levels of management. For example, the ability to develop application programs may be a function which management prefers to control and to make available only to selected personnel. This would require that application-development software, such as compilers, be controlled and be accessible only to an authorized user group(s). In other cases, the program functions associated with the manipulation of corporate financial data may require tight control to avoid misuse by unauthorized personnel. In this example, perhaps only certain people in the financial organization and some of the auditor's staff would be allowed to execute those functions. This pattern of control would affect system design and must therefore be defined during data analysis.

Many of the patterns which relate to management requirements may have been defined during one of the other pattern-analysis processes. This section is included as a reminder that management may have requirements which generate more complex patterns than other system users. Extra care in searching for those patterns is therefore necessary.

PATTERNS OF CHANGE

During data collection the functions and data elements of the system which can be expected to change were identified (see Figures 5-3 and 5-5). These must be grouped into patterns of stable functions, stable data elements, functions likely to be changed, and data elements likely to be changed. In the case of functions and data elements which are likely to change, additional patterns are often discernible, and the source of expected change may be a significant factor in defining those patterns. For example, if a specific set of functions or methods is expected to change because of pending legislation, those functions form an identifiable pattern. If union negotiations make it likely that hourly wage earners will receive a new contract, their wage rates and benefits form another pattern because of that potential change.

The sources of change in a specific organization may be extremely varied, and it is probably impractical to identify all of them as patterns to be considered during system design. However, any patterns of change—especially significant change—which can be predefined will

---

APPLICATION: PAYROLL AND PERSONNEL

1. Contract negotiation—May 1983
   1.1 Hourly workers in the factory and warehouse
       1.1.1 Pay rates
       1.1.2 Work rules

Significant changes in work rules are expected as a result of this particular negotiation process

2. Reorganization—September 1983
   2.1 Salaried workers
       2.1.1 Marketing Division
       2.1.2 Manufacturing Division
       2.1.3 Corporate Headquarters

Executive management is considering a significant reassignment of functions among the three groups listed. This could result in the following changes:

- Divisional and departmental identifiers
- Shifts of personnel between organizations
- Changes for specific individuals:
  —Job title
  —Organizational reporting level
  —Salary

Figure 7-6   Patterns of change.

aid in formulating the design. It is, of course, appropriate to consider all parts of a system subject to change, but, realistically, some parts are more likely to change than are others. This analysis of change patterns should help to define the more likely areas. An example of how selected patterns of change can be documented is shown in Figure 7-6.

SUMMARY

The patterns defined for analysis in this chapter are those which are found in almost every information system; however, these are not the only possible patterns of interest. Each system has unique relationships, based on the work flow, organizational structure, management style, and so on, of the organization. The system designers must therefore search for relationships which will be significant in designing the information system. Each such pattern must be defined and documented during this data-analysis phase.

Many patterns which are detected in this process will prove to be repetitive and/or overlapping. That is, the analysis of user-group patterns will uncover patterns in data-access requirements as well as geographical-location patterns. It is also possible that analysis of geographical patterns will uncover previously unrecognized user groupings. It may be quite difficult to clearly document the patterns which are defined, because of these interrelationships and the overlap among multiple patterns. (The figures provided earlier in this chapter show how many relationships typically exist among patterns.)

One possible way to document these patterns is to describe each graphically on a separate sheet, which can then be converted to a transparency. If the patterns are carefully drawn, related transparencies can then be placed on top of one another to show the overlapping patterns and relationships. In a complex system with many levels of interrelated patterns, it may not be workable to overlay all patterns simultaneously, but various combinations of patterns can be compared and analyzed together. In this way all patterns and their relationships can be analyzed gradually but thoroughly.

8

RELATING PATTERNS
TO SYSTEM DESIGN

Each type of pattern defined in Chapter 7 can be related to one or more specific aspects of system design, as described in this chapter. A general ground rule for good system design is to partition according to patterns. This means that functions, applications, computer facilities, databases, and networks can be partitioned (in a centralized or a distributed system) to match the patterns of relationships which exist in the organization(s) to which the information system is related.

It is important to distinguish between stable patterns which are unlikely to change and therefore form a good base for system design, and patterns which fluctuate over time and thus may be poor choices on which to base system-design decisions. The following discussion of how patterns relate to system-design choices is arranged in the same sequence as the discussion of pattern recognition in Chapter 7.

USER-GROUP RELATIONSHIPS

The ability to meet users' processing and data-access requirements is the most important aspect of an information system. User-group relationships are therefore an essential element in the design of that system. Chapter 7 discusses various ways to group users—by functional requirement patterns, by data-access patterns, and/or by geographical patterns. Each pattern of this type can help to define the system design.

Functional-requirement patterns, which group users by common functional needs, usually define how application logic can be partitioned to form programs or related sets of programs. The situation to be avoided in structuring application programs is the one shown in Figure 8-1. In this example, overlapping or common functions for user group A and user group B have been implemented in one or more common applica-

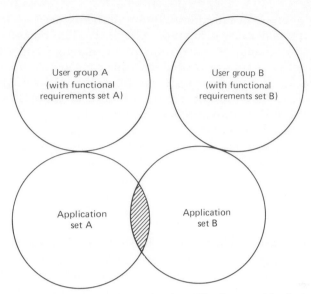

Figure 8-1   User groups and application partitioning.

tion programs, indicated by the shaded area in Figure 8-1. Partitioning application logic in this way increases vulnerability to change. If user group B's requirements change, for example, but the corresponding requirements for user group A do not, the application logic must be not only changed but also repartitioned to match the changed relationship between the two user groups' requirements.

If a situation similar to that shown in Figure 8-1 exists, the following steps are appropriate. First, reevaluate the patterns formed by analyzing the requirements of the user groups. It is possible that there are really three rather than two user groups, each with a unique set of requirements. In this case, repartitioning of the application functions to match those new user groups is in order. It is also possible that the first definition was correct; i.e., that there are two groups of users who share some common requirements. In this case, the correct solution is generally to duplicate the shared functions in the set of application programs which serves each user group.

Designers and programmers often resist the duplication (or replication) of application logic for two valid reasons. One is the possibility that performance will be lower than with one copy of the logic; the other—more serious—is the potential difficulty of keeping multiple copies in synchronization when changes occur. The potential performance degradation is clearly a manageable problem on contemporary computer systems. In fact, multiple copies of the same code may improve performance, not degrade it, since multiple simultaneous accesses to the same program copy may cause contention and delay. The potential problem associated with change is more difficult to manage, but with proper documentation and control procedures this can be accomplished. The management of change is essentially no different in this situation than in the general case of controlling program modifications and interactions, even when there is no duplication of program logic.

Clearly each situation of this type must be evaluated on an individual basis. In some cases it may be acceptable for two application sets to share common logic, especially if the shared functions are relatively stable and will probably continue to be required by both sets of users. The common logic must, however, be identified and segmented into separate programs, modules, or subroutines to facilitate change or partitioning if necessary.

Data-access patterns, which group users by their need for access to particular sets of stored data, may determine how the data managed by the information system will be partitioned. If the total set of data elements to be controlled by the system is considered to form a system database, it may be necessary to segment the system database to produce a more manageable set of databases and/or files.

If all data are to be stored at a central location, partitioning may take the form of separating data elements into files, independent databases, or areas within a single database. If a decision is made to use a distributed-system structure, as discussed in Chapter 10, the system database may be partitioned and/or replicated to form a distributed database. In that case, data-access patterns may help to decide which data elements belong in each part of the distributed database. User-access patterns are one of several factors used to decide how to organize the system database—in terms of structuring a single database and/or of segmenting a distributed database. Other aspects of those decisions are discussed later in this chapter, and the process of defining the database structure is covered in Chapter 11.

Geographical patterns, formed by the location of system users, can help to define the ways in which application functions and data storage are partitioned. Those patterns can also help to define where information-processing components, and possibly also data-storage devices, are physically located. The allocation of functions and equipment geographically is discussed in a later section of this chapter.

DATA-ACCESS NEEDS

The data-centered theory of system design places database content, structure, and use at the top of the priority list in system analysis. Not everyone agrees with this ranking, but the database (or set of databases) is certainly a focal point and a major element in most contemporary information systems. Patterns of data access are therefore important in formulating the system design.

The preceding section discusses one type of data-access pattern, that formed by grouping users who share the need to access certain types of data elements. Two other patterns are potentially important; one defines when and how access to specific data elements occurs, and the other separates production-mode access from *ad hoc* access.

Access-time and -type patterns group data elements which are accessed interactively from those which are not and also define the periods of time in which accesses occur. Those patterns can help to separate data elements into some or all of these categories:

- Data elements which are accessed interactively, with fast-response requirements, at all times (24 hours a day, 5 or more days a week)

- Those which are accessed interactively, with fast-response requirements, during specified time periods (usually the business day of roughly 8 to 12 hours, possibly only 5 days a week)

- Those which are accessed in batch mode, during specified periods daily

- Those which are accessed, in batch mode or interactively, at weekly, monthly, quarterly, or other specified intervals

Many data elements fall into multiple groups in the overlapping patterns which typically result from the above categorization. Data elements must be grouped, for system-design purposes, in terms of the most rapid access requirements. These patterns will be useful during database design, as described in Chapter 11. Data elements partitioned by access patterns can be placed on different types of storage devices and/or in different segments of a distributed database. For example, data elements accessed frequently may be placed on fast-access disk devices located geographically as close to the users as is practical. Data elements accessed only in batch mode may be stored on slower-access disks or, if access is infrequent, on off-line or archival-storage devices. These elements may also be grouped at locations geographically remote from the users.

Another possible method of partitioning, which is not used as frequently today as it was in the past, is to separate data elements which are accessed for inquiry only, at least during some time periods and/or by certain users, from those accessed for update. If this type of pattern exists, it can be used to segregate data into different files, areas, or databases. It may also be used to identify data elements which can be copied for inquiry-only use, while the original version may be updated in parallel. The inquiry-only copy or copies may form a distributed replicated database (see Chapter 11).

Production- and ad hoc-access patterns separate data elements which are accessed routinely from those needed for access in ways which are not predefined. As in many patterns, this will usually cause a great deal of overlap, and many data elements will belong to both groups.

Data elements accessed in production mode can be further separated by access time and type, as discussed earlier. Special attention is needed for data elements which are expected to be accessed via query or other ad hoc facilities. These data may require storage on fast-access devices to provide the quick response needed; they may also require very flexible data structures. These topics are discussed in Chapter 11.

Data elements which fall into both the production-access and ad hoc-access groups must be analyzed carefully. The need for ad hoc access may, in some cases, dictate the need for rapid access even if quick response to the same data elements is not needed in production use. This might be the case, for example, when financial data are used by analysts or executives in making investment decisions (which may call for the very rapid initiation of electronic funds transfers). In other cases, ad hoc access does not imply rapid response. For example, researchers in a library may need to browse through catalogs and/or document excerpts. Rapid access is neither financially justified nor a user requirement in such a situation.

GEOGRAPHICAL GROUPINGS

The geographical patterns defined in Chapter 7 affect many aspects of system design. As might be expected, these patterns help to define where facilities and functions will be located.

User locations define where terminals or work stations* will be placed. An initial assumption is generally made that each user will require a terminal, but as the design progresses and cost trade-offs are made, it may be determined that a group of colocated users can be served by fewer terminals than there are users. Of course the terminal or work station devices must be located where the users can conveniently access them.

The combination of patterns is also important. If a geographical cluster of users is also a user group with shared functional requirements, this combined pattern has the potential to define a location for distributed

* A work station is a multifunction terminal or set of terminal equipment.

functionality. If those users also share data-access requirements, the potential exists for not only distributed processing but a distributed database to serve that cluster of users.

Users—such as traveling salespeople—who will access the information system from different locations at different times usually define the need for dial-in access facilities. This pattern may also indicate the need for specialized devices such as portable terminals or for access procedures which can use generally available facilities such as telephones.

All information about the locations of users, whether fixed or mobile, is also input which will be used during network design (see Chapter 12).

Computer-equipment locations and locations which may be suitable for the installation of equipment help define where the computer-based functions of the new system will be carried out. Existing facilities may define locations suitable for the allocation of information-processing functions, database facilities, and/or network-processing functions. Locations suitable for new equipment may also offer any or all of these possibilities. The information about current and potential future equipment locations will be used in network design (see Chapter 12).

Communications-facilities locations, whether used for voice, data, or combinations, also supply input to network design. The relationships among user locations, computer-equipment locations, and communication-facilities locations also form an important factor in network design.

INTEGRITY-LEVEL
REQUIREMENT PATTERNS

The grouping of data elements in terms of requirements for integrity protection separates those elements which require a high level of correctness from those in which a greater degree of error can be tolerated. Those patterns define the types of integrity-protection and validation routines to be used. For example, certain data elements may require specialized input methods and strict correctness-validation routines. This may lead to the development of terminal-user procedures different from those which are adequate for other types of data with less stringent integrity requirements. The protection of data stored in the information system may also take different forms, depending on the integrity-level requirements.

Patterns of this type do not exist in every system; in some applications it is impossible or impractical to define data elements which can be given less than the highest level of integrity protection. However, if there are elements which can be handled in this way, the integrity-protection patterns may indicate ways to partition the total system database. In fact, it may be impractical to implement different levels of integrity protection unless the different classes of data can be segregated into separate files, database areas, databases, or partitions in a distributed database.

SECURITY- AND/OR PRIVACY-REQUIREMENT PATTERNS

The patterns of security protection and privacy protection defined in the process described in Chapter 7 are used during system design to separate data and functions which require different levels of protection. Patterns are especially important in this context, because partitioning is the most basic method used to provide security or privacy protection. Typically, data and functions which require strict protection are segregated from those which need less stringent safeguards, and in some cases the segregation process itself improves the level of protection.

Each protection level to be provided in the system may define the need for separate facilities. For example, the requirement for a high level of security protection might imply the need for separate terminals in protected locations, to be used only for secure data input and output. Similarly, there may be a need to configure separate network links for secure and insecure data and to provide encryption mechanisms on the secure links. The patterns of protection therefore usually dictate how programs, data elements, information-processing components, network-processing facilities, and terminal devices will be partitioned so that an appropriate level of protection can be provided.

The relationship between patterns of security/privacy protection and other patterns—especially user-group relationships and data-access needs—is also important. If the patterns of data access are exactly the same as data-protection patterns, the data-partitioning process during system design is quite straightforward. In other cases, data-access patterns and/or user-group patterns may define one method of partitioning and protection requirements may indicate a different method. In that case, a trade-off must be made between the conflicting patterns. In systems which place emphasis on security and/or privacy protection, the data-protection needs are typically the determining factor in design decisions. In systems with less severe requirements for security and/or privacy protection, partitioning is usually done on the basis of access patterns.

Privacy protection applies only to data stored or processed by the computer-based system. The emphasis in security protection is typically also on data, but in many systems functions and computer equipment also require protection. Many of the requirements and techniques for the protection of functions and equipment are internal to the system and are first considered during system design. Some protection requirements may, however, be defined during data collection, and those requirements may form patterns which mandate methods of functional partitioning.

MANAGEMENT-CONTROL PATTERNS

The patterns defined in this category, such as management data-access patterns and functional-requirement patterns, will generally also be described in other categories. Patterns of management control may also be defined, based on the information concerning management philosophy which was obtained during the data-collection process in Chapter 5. Management-control patterns potentially define the following:

- Functional partitions

- Data partitions

- Allocation of control functions

Functional partitioning in this context may involve separating functions which require tight management control from those which do not. Sometimes the functions involved are applications; in other cases they are functions such as program-development tools (i.e., compilers, debug facilities, etc.). Partitioning may also separate functions so that those controlled by each level of management are identified; e.g., functions of interest only to executives may be segregated from those of interest to middle-level management, and so on. The reasons for partitioning may be to locate functions close to the users of those functions and/or to apply different levels of security control.

Data partitioning in terms of management requirements is exactly the same as functional partitioning. Data may be partitioned to place elements close to the management users of those elements and/or to apply different levels of security protection.

Allocation of control functions is somewhat more complex and relates very closely to management style. Those patterns help define how to control the use and change of functions and/or data elements. (This is one specific aspect of security protection and control.) Management-control patterns may also indicate how functions and/or data are partitioned. For example, functions and/or data over which management wishes to exercise tight control may be separated from other functions and/or data. The functions and data which are to be tightly controlled are usually placed at a central location, while other functions and data may be distributed based on other design criteria. Management-control issues may therefore determine where computer equipment and stored data are physically located.

PATTERNS OF CHANGE

The identification of elements of the system and its environment which are expected to change—or to change more rapidly than other elements—is important in system design. Patterns of change may identify either data elements or functions for which methods of change must be specified. They may also identify partitioning methods for data or functional elements.

Although it is not always necessary to partition data elements to be changed frequently from those which are not likely to change, this may be of considerable benefit. If change can be isolated to specific records, files, database areas, or other segments of the total system database, it will be easier to isolate the potential impact of change. The relationships among patterns must of course be evaluated. If patterns of change are dramatically different from patterns of data access and/or from patterns of security-/privacy-protection requirements, the patterns of change are usually considered secondary, and data partitioning will more likely be done in accordance with other defined pattern(s). In other cases all the patterns may be consistent, then the partitioning decisions can be made easily.

Patterns of functional change can help to define how functions are partitioned into application programs, modules, and/or subroutines. Separating functions which change rapidly from those which do not will help to ease change and can help to ensure that changed functions can be tested and installed without affecting other functions. As in the case of data change, isolating frequently changed functions is not absolutely essential but will usually improve the system's flexibility. Techniques to be used in designing for flexibility and adaptability are covered in Chapter 13.

SUMMARY

The patterns discussed in Chapter 7 and in this chapter may not include all those found when studying a specific environment. If other patterns are identified during data analysis, their relationship to system design must be defined. As the discussions in this chapter point out, patterns are most often used to partition functions and/or data during system design. Partitioning is a major activity during system design: functions must be partitioned into applications, programs, subprograms, subroutines, and modules; data elements must be partitioned into files, subfiles, records, database areas, database partitions, or separate databases. Using patterns identified by analyzing data collection to describe the system requirements and environment can help to ensure that the system design accurately reflects reality.

It is extremely important to *recognize* patterns which exist in the environment or which will exist in the future as the environment changes. It is possible, when studying system requirements, to *imagine* patterns which would be desirable and/or to impose desirable patterns on the data collected earlier. A system designer may fall into this trap if an inadequate amount of data is collected or if the designer allows his or her prejudices to intrude into the data collection and analysis process. Objectivity, while difficult to achieve, is essential to all aspects of system design.

Design decision	*Related patterns*

1. Partitioning application functions into:
 	
  • Programs
  • Subprograms
  • Modules
	1.1 Users' functional-requirement patterns
	1.2 Users' geographical-location patterns
	1.3 Security-protection patterns
	1.4 Management-control patterns
	1.5 Patterns of change

2. Partitioning stored data into:
 	
  • Files
  • Databases
  • Distributed database segments
	2.1 Users' data-access patterns
	2.2 Users' geographical-location patterns
	2.3 Production- vs. ad hoc-access patterns
	2.4 Integrity-level requirement patterns
	2.5 Security-protection patterns
	2.6 Privacy-protection patterns
	2.7 Management-control patterns
	2.8 Patterns of change
	2.9 Access-time and -type patterns

3. Partitioning stored data onto different speed devices	3.1 Access-time and -type patterns
4. Stored data protection methods	4.1 Integrity-level requirement patterns 4.2 Security-protection patterns 4.3 Privacy-protection patterns
5. Function and equipment (computer, terminal, and network facilities) protection methods	5.1 Security-protection patterns

Figure 8-2 Relating patterns to system design.

Design decision	Related patterns
6. Selecting locations for terminals	6.1 Users' geographical-location patterns 6.2 Types of locations patterns (e.g., office environment, factory environment, etc.)
7. Selecting locations for computer equipment	7.1 Users' geographical-location patterns 7.2 Users' functional-requirement patterns 7.3 Types of location patterns 7.4 Management-control patterns
8. Selecting locations for stored data: • Files • Databases • Distributed database segments	8.1 Users' geographical-location patterns 8.2 Users' data-access patterns 8.3 Types of location patterns 8.4 Management-control patterns

9. Designing the network	9.1 Geographical-location patterns:
	    9.1.1 Users' terminal locations
	    9.1.2 Computer-equipment locations
	    9.1.3 Existing network facilities
	9.2 Traveling user patterns

10. Designing terminal-user interfaces	10.1 Integrity-level requirement patterns 10.2 Security-protection patterns 10.3 Privacy-protection patterns
11. Allocating control functions	11.1 Management-control patterns 11.2 Geographical-location patterns

Figure 8-2 (Continued)

Figure 8-2 shows in matrix form how the patterns most frequently found during data analysis are related to the various aspects of system design. This matrix can be used during the pattern-detection process described in Chapter 7, and also during the system-design process described in Section 5, as a checklist to ensure that all potential patterns have been analyzed and correctly related to the different steps in the system-design process.

SYSTEM DESIGN

9

DESIGNING
USER INTERFACES

The first, and in many ways the most important, step in system design is to define exactly how the users will interact with the sytem. The definition of user interfaces must include both interactive methods, which are of primary importance in most systems today, and batch methods. Since this book concentrates on the high-level or strategic system-design issues, this chapter does not provide a complete description of how to design user interfaces. However, even at the strategic level it is necessary to place a great deal of emphasis on interactive interfaces, because they are a major factor in the success of any information system. Batch interfaces, which consist of report formats, can be defined in detail later, during the detailed system-design process.

It is sometimes argued that user interfaces cannot be defined until late in the system-design process, when all the strategic-design decisions have been made. For example, decisions such as whether the system will be centralized or distributed and whether the database will have a network or a relational structure are part of the strategic system design. It is true that there are relationships between those decisions and the system's user interfaces; however, the interface design ought to be based on the users' needs and working environment, not on the information-system components or structure. In some of the most successful systems the user interfaces had been defined—in a great deal of detail—before any system-design decisions were made and before the computer and terminal vendors were selected.

The first step in the design process is to determine which interfaces will be interactive and which will be in batch mode. Then it is necessary to decide which interfaces will be predefined for routine, production-mode operations and which will be ad hoc requests for queries and reports. Most of these decisions will have been made, at least tentatively, as a result of the data-collection process described in Chapter 5. Data collection includes the definition of output requirements, including which types of output are required interactively and which in batch mode, and also the definition of system-usage modes: interactive, pre-defined, ad hoc, etc.

At this point in the system-design process, tentative decisions made during data collection must be reevaluated. After completing data collection and analysis, the system designers ought to have a good understanding of the users' requirements and working environment. Firm decisions can now be made concerning system-usage modes and input/output requirements. The focus during data collection was on defining output requirements, since those are the end-product of a computer-based information system. During the design of user interfaces, the focus shifts to input formats and methods, with secondary emphasis on output.

It is essential that the design of the user interfaces be a cooperative process between the designers and the prospective users. If the users have been successfully enlisted in the system-design effort during data collection, they can be very helpful at this time. If good relationships have not been established earlier, that must be the first priority when beginning the design of user interfaces.

INTERFACE DEVICES AND METHODS

There are many different types of interface devices, most of which are used for interactive input and/or output. A brief review of available devices and techniques is appropriate before discussing how to select the device(s) and method(s) to be used. The following classes of devices, each of which is described briefly in this section, can be considered:

- Video-display terminals (also called CRT for cathode-ray tube)
- Special-purpose terminals
- Hard-copy printers
- Voice input and/or output devices
- Viewdata

VIDEO-DISPLAY TERMINALS

Video-display terminals (VDT) are probably the most commonly used devices for interactive work. As a minimum, a VDT consists of a display screen and keyboard. The display may have graphics capability, to produce bar charts, pie charts, and similar business graphics, and/or to handle complex character sets such as Chinese or Japanese. The VDT screen may be able to display multiple colors, which are often used in conjunction with graphics. The VDT keyboard usually includes multiple function keys, each of which signifies some command or combination of input characters when activated.

Many other options are available with some VDT models. A hardcopy printer may be offered, either as an adjunct to each display or, more often, as a device shared among several displays in a cluster. A light pen may be optionally provided, or the display screen may be touch-sensitive; in either case selections or choices can be made by pointing to the desired data displayed on the screen. Other possible options include a badge-reading or card-reading device for the entry of data stored in encoded form on a badge or card. A document-reader option may also be available. Badge readers, card readers, and document readers are used either to enter user-identification data or to enter constant data to avoid key entry.

Video-display terminals are used in several ways, which fall into these three classes:

- Forms mode

- Tutorial prompting, or menu mode

- Natural-language mode

Forms mode is probably the most common usage method, especially for transaction-processing applications. Forms mode is sometimes called "fill in the blanks," because a form is displayed on the screen so that the user can fill in the required data in the blank spaces on the form. The input sequence begins when the terminal user depresses a function key associated with the type of transaction to be entered—or in some cases selects a choice from a menu displayed on the screen. The system responds by displaying a form which contains all the data items necessary to complete the transaction. For example, as shown in Figure 9-1, an order-entry application might involve entering the customer's identifying number, the desired shipping date, and the identifying number and quantity of each item ordered. The terminal cursor is positioned at the beginning of the first field to be entered, and each field may be edited for correct size, format, range, etc., as it is entered. Usually only the data fields are transmitted to the processing computer; the form information is discarded. Output may be inserted in the same form, or the screen may be cleared and an appropriate output message displayed. In the example shown, output would probably include the customer name and address,

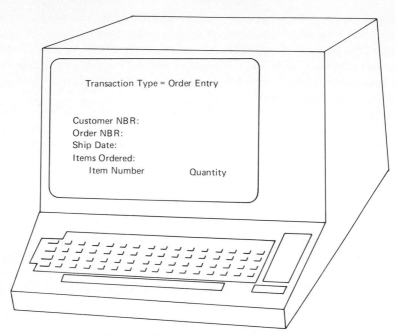

Figure 9-1 Forms-mode user interface.

obtained from the database and displayed below the customer number. This would allow the user to verify that the correct customer record has been retrieved for processing.

Forms mode is often the easiest, and potentially the most accurate, method of interaction for users who have keying skills or can acquire them easily. (The number of people with some level of keying skills is increasing, as many jobs now require the use of terminals.) Depending on the system design, forms mode may impose a relatively high level of overhead on the computing system. This is especially true when a large information-processing system does all the forms handling as well as the interactive editing of data during entry. In many systems today, the storage and management of terminal forms and the editing of data during entry are handled by the terminal controller or by a satellite processor close to the terminal devices. This minimizes the load on the central system, if any, and also reduces the load on the communications links; the result may be to lower the cost of communications. These system-design factors must be taken into account when deciding if forms mode will be used.

Tutorial prompting, often called *menu mode*, is the second interface method. In this method the user is given a set of choices, or menu, from which the desired selection is made. Each selection in turn leads to a new set of choices, as shown in the example in Figure 9-2 of the beginning of a database-query sequence.

All the necessary data may be supplied by means of the choices made, or the final choice may lead to the display of a form which is then filled in as described earlier. Tutorial mode makes it very simple to use the terminal, as the person entering data need know very little about the computer-based system or, in some cases, even about the application. All valid choices are displayed, so that editing of the input supplied by the user is quite simple.

Menu-based interfaces can also minimize the amount of data which must be keyed in. However, this mode can be extremely boring for the experienced user and can decelerate the entire operation because of the wordiness and slow pace of the selection method. This interface is best suited to ad hoc queries and reports, which can be made very flexible, and/or for situations in which users are typically inexperienced and need constant training even while performing production-mode data entry. (If the user turnover rate is high, users may need more prompting than with a low turnover rate.) This mode permits the immediate editing of data elements, just as forms mode does.

If menu mode will be used repetitively by experienced users, a "fast-track" option must be provided. This allows an experienced user to bypass most or all of the menus and prompts and simply enter the required data. A system which combines full tutorial prompting and a fast-track option can provide the best of both approaches. An inexperienced user, a person who uses the facility infrequently, or a user being trained can access the menu interfaces, while an experienced user or one

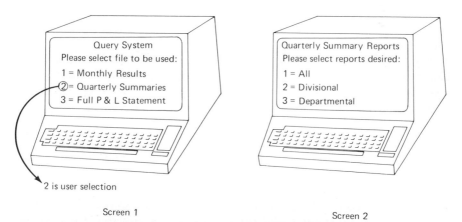

Figure 9-2   Menu-mode user interface.

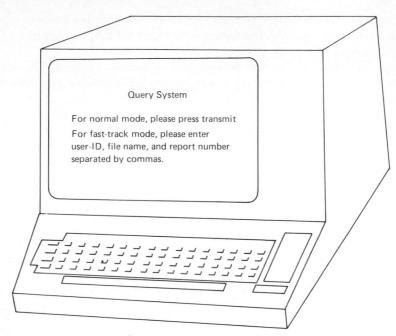

Figure 9-3 Menu-mode fast-track option.

entering production data can choose the fast track, both simultaneously on the same system. An example of a fast-track option is shown in Figure 9-3.

Natural-language mode is a third method of using VDT devices. The term "natural language," in this context, means that the input is as close as practical to language as normally spoken or written. The language used is the one which is natural to the user; e.g., English, French, Japanese, and so on. There are two problems associated with this mode of input which severely limit its usefulness. One problem is that many users, especially managers and professionals, are not sufficiently at ease with keyboards to wish to type the large quantities of text which are typically required to use natural language. The other problem is that it is extremely difficult technically to process natural-language input and determine its meaning.

The difficulty of evaluating language syntax and semantics is a problem which has been studied for many years—so far with limited success. Natural language is very complex and ambiguous, so much so that programming languages have been artificially simplified so that their syntax and semantics are regular and unambiguous enough that they can be analyzed by a computer. Natural language, in contrast, is extremely irregular and rich in forms and nuances. Advances are being made in the processing of natural language; however, at the present time it is necessary to limit the vocabulary and semantic forms permitted for computer input. Within well-defined boundaries this type of seminatural language can be used today. An example of this type of input, with associated output, is shown in Figure 9-4.

The optional features available with some display terminals can also be used to supply input. A badge-reading device can provide user identification and/or specific data elements. Most often, a badge reader is used to identify the person who is accessing the terminal, through encoded identifying information on the badge, thus eliminating the need to enter identifying data via the keyboard. Of course this type of identification is far from foolproof, as the only thing positively identified is the badge, not the person using it. A badge or card reader can also be used to enter information concerning jobs, particularly in manufacturing operations, as discussed in the section on special-purpose terminals which follows. Optional document readers associated with some VDTs are used in similar ways.

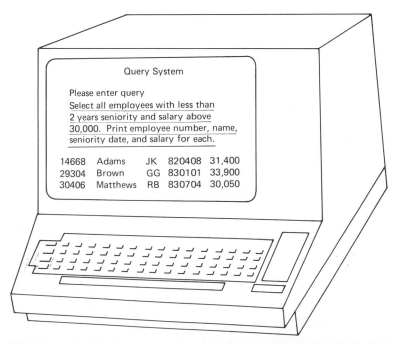

Figure 9-4   Natural-language user interface (user input underlined).

SPECIAL-PURPOSE TERMINALS

Special-purpose terminals are available to serve many purposes, the most common of which are in the applications associated with banking, ticketing, manufacturing, and point of sale.

In banking, specially designed terminals are available for both in-bank use by tellers and use by customers. A teller terminal usually consists of a small display and keyboard and may include a magnetic-stripe reader which can be used to enter data from the customer's credit card or bank card. There may also be a small auxiliary keyboard through which the customer can enter a personal-identifier number (PIN) for positive identification.

Customer-use terminals are commonly called ATMs, or *automated teller machines*. An ATM includes a display, keyboard, and stripe-card reader, as well as a cash-dispenser and deposit-accepting mechanism. The intent for these devices is to minimize keying and provide an extremely easy-to-use interface. The ATM usually displays prompting input, and the customer must enter his or her bank card as an identifier and a key to the account to be debited or credited. A PIN is also required to further identify the customer. Human engineering is the key to the success of ATM devices, many of which are on the market today. An example ATM device is shown in the Figure 9-5 photograph.

Terminals which issue tickets are found in airline, train, subway, and entertainment applications. Ticketing terminals are generally designed for use by customers. The customer follows a simple dialogue with the terminal via the display screen, inserting a credit card into a reader on request, so that the ticket can be charged automatically to that account. The output provided is a ticket printed for the desired trip or entertainment function. Ticketing terminals are also used by the airlines behind the reservation counters, both to issue tickets and to prepare boarding passes.

Many different types of special-purpose terminals are used in manufacturing applications. Most use some form of badge or card input to minimize the amount of keying required. In many factories an encoded card, badge, or ticket accompanies each job or item as it moves through the production process. When a worker completes a specific operation, the card, badge, or ticket for that job is entered into a reader, usually accompanied by the worker's badge, indicating that the person has completed a specific step in the manufacturing process for that job. A shop-floor terminal of this type is shown in the photograph in Figure 9-6.

In some manufacturing applications, wand readers or scanners are used to mechanically record part numbers and similar data as parts and assemblies move from place to place. In some large-scale manufacturing processes, robots are increasingly used for more complete automation of

Figure 9-5 Automated-teller machine.

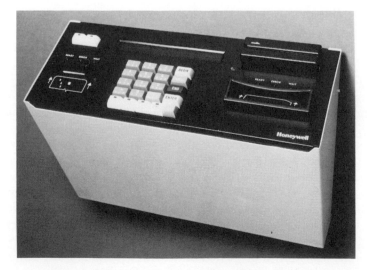

Figure 9-6 Manufacturing shop-floor terminal.

selected tasks. Robotics, because of the complexity of the subject, is outside the scope of this volume.

Terminals for point-of-sale applications, most of which make use of some form of magnetic-reading device, are coming into widespread use. In supermarkets, for example, a scanner which reads the universal product code (UPC) markings included on many grocery items can be installed in the checkout counter. The improved speed and accuracy of the checkout procedure in these semiautomated systems can be very impressive. There have been several attempts to interface point-of-sale applications, especially in supermarkets, with on-line-banking systems for automatic check verification and potentially for electronic funds transfer. Although the success to date in those experiments has been limited, it is probably only a matter of time before this type of connection becomes commonplace.

There are other types of special-purpose terminals, although none in as widespread use as those described above. If an application is large enough, it may justify an investment in custom-built devices to meet the specific requirements of the organization. For example, the McDonald's fast food chain purchased specially designed terminal systems from ITT Courier to assist in speeding up the customer-payment process. This application, which requires tens of thousands of terminal devices, is large enough to make custom design attractive. The potential payback in increased efficiency and accuracy, often with inexperienced teenagers as terminal users, was an important factor in the investment decision.

PRINTERS

Printers of many types are used as output devices, sometimes with an associated keyboard for input. It is convenient to classify printers as either hard-copy terminals or batch devices.

Hard-copy terminals are very similar to VDTs in that they provide a keyboard for data entry but use the printer as the output device (and also to record the data which have been input). Interactive terminals were originally hard-copy devices, but today VDTs have almost completely replaced those devices. Hard-copy terminals have the disadvantage of being slower and noisier than VDTs, and their only offsetting advantage is that they automatically provide a hard-copy record of the data which have been input and output. Most often, a printer is provided either as an optional feature of a VDT or as a shared device serving a cluster of VDTs. In this situation, the printer is used for logging data which must be retained in hard-copy form and also for the output of data, such as documents, which must be passed on to other people. Many office-based work stations include letter-quality printers, which can create smooth-form documents in multiple fonts, with justified margins and proportional spacing. Some printers have the ability to generate multiple character sets, so that complex languages such as Japanese and Chinese can be processed.

Batch printers can be part of a remote-job-entry terminal or part of the configuration of an information-processing system—either a host computer or a satellite processor. Batch printers are of two types, impact printers and nonimpact printers. The former are typically limited to speeds of about 1500 lines per minute; the latter can operate at up to 18,000 lines per minute. A complete survey of printer technology is outside the scope of this discussion, and the batch printers are included only to complete the list of user-interface devices.

VOICE INPUT AND/OR OUTPUT DEVICES

Voice input and/or output devices are now available at a sufficiently low cost to justify their use in some applications. Voice output has been widely used, especially in banking. Some ATM devices, for example, use voice output to guide the customer through the steps necessary to use the terminal. Account-balance inquiries via Touch-Tone telephone, with voice response from a central computer system, are commonplace in many banks. One of the advantages of voice response is that telephones can be used as low-cost, readily available terminals.

Some voice-output devices use words or phrases recorded digitally and stored; these can be combined to produce the required messages. Other devices use voice synthesis. In this technique, basic language sounds—called phonemes—are combined to produce a theoretically unlimited vocabulary of output messages. In practice, combining phonemes into natural-sounding words and phrases is quite an art, so that even the best of today's voice-output applications using this technique tend to sound mechanical. For small amounts of output this is acceptable, but it can be irritating if long voice messages are generated.

Other voice-output techniques include linear predictive coding, which generates output using a mathematical model of the human throat. Although theoretically flexible, output quality is low and the sound unnatural. Still another method combines linear predictive coding with voice-waveform encoding; this shows considerable promise but is not in commercial use. In each case, the choice of a voice-output method must be based on a trade-off between the output quality (naturalness) and computer-resource costs; with today's devices, higher quality usually means higher cost.

Many people believe that voice input will represent an essential breakthrough in computer use, largely because input devices rely heavily on some form of keyed input at present. Speaking is, for most people, a much more natural activity than keying, and they can therefore be much more comfortable with computer systems which accept voice input. However, the technical difficulties associated with voice input are much greater than those associated with voice output. Those difficulties are of two types, one concerned with recognizing what the speaker is saying and the other with evaluating the meaning of the spoken material.

Techniques for speech recognition have advanced enormously but still have some distance to go before it is practical to use voice input in all cases. Devices today can usually recognize only a limited number of words, which represent the working vocabulary. Many devices also require that each person who is to use the device first speak a predefined phrase one or more times to "train" the device before use. This allows the device to compare the waveforms generated by that speaker's voice with a predefined set for the phrase and adjust its algorithms so that it will recognize input spoken by that person.

Several organizations, including Bell Laboratories, have been working intensively on voice-recognition devices which do not require training and can adjust heuristically to each new speaker. Of course it is technically difficult to operate in this mode and still keep the error rate within acceptable limits. Errors in voice recognition are cases of either unrecognizable input or input which is recognized as the wrong word or syllable. The latter is the worst possible error, and the potential for this type of error exists in all voice-recognition devices. To minimize the possibility of misrecognition, many voice-input devices display or speak the input back to the user.

Many factors have a potential impact on the level of errors and misrecognized words. Variations in the speaker's voice because of illness, emotion, and other factors may cause errors. (These factors theoretically have no effect on the voice waveforms, but in practice such effects are noted.) A noisy environment is a potential problem, as the voice-recognition device may receive not only the speaker's input but also interference from other sources. However, it is illustrative of how the technology has advanced that General Electric was reported in 1981 to be planning the use of voice-input devices in some of its factory locations. In the planned application, the worker will speak the required input data into a microphone suspended around the neck, while using both hands in the manufacturing process. The input methods which voice input will replace require the worker to stop processing periodically to use a key-entry or badge-entry device. Tests made in the factory by G.E. were apparently very encouraging, and they believe that higher levels of accuracy can be attained with voice input than with other methods.

The other technical difficulty associated with voice input is how to determine what the input means, assuming that the syllables and words have been correctly recognized. Systems today limit the vocabulary used, so that the computer which evaluates the input has a relatively small number of possibilities to consider. In a factory parts-inventory application, for example, the user typically speaks a part number and a quantity for each transaction. Only digits are involved in this type of application, and the computer has only to determine if the correct number of digits was entered and if the digits which form the part identifier match a known identifier. Many present-day applications avoid the use of words and permit only digits as input, to simplify the voice-processing problem.

The ideal voice-input device would accept free-form natural language from any speaker and correctly evaluate it. Assuming that the problem of recognizing what the speaker says is solved, determining the meaning of natural-language input remains. This problem was discussed earlier, in the context of VDT input, and is identical whether the input is voice or keyed. It is difficult to analyze the syntax and semantics of truly free-form language; in fact, in the present state of the art this is beyond the capability of computer-based systems. Applications today must, therefore, limit the vocabulary and syntax used so that the input can be recognized and processed correctly.

There is one exception to this general limitation in the use of free-form spoken input. Voice mail, one of the variations of electronic mail, handles messages by recording spoken input, storing it, and later replaying the message to the designated recipient(s). In some cases the recorded message may also be transmitted between computers before it is replayed for the addressee. This application does not require any analysis of the voice input and is therefore well within the state of the art.

VIEWDATA

Viewdata is used here as a generic term covering a variety of possibilities for data input and/or output using television sets. Many terms are used to describe this technique; some are specific, such as the Prestel system in the United Kingdom, the Antiope system in France, and so on; others are generic, such as the terms viewdata, teletext, and videotext. Viewdata, to use a single name to refer to all these forms, represents a major new possibility in the 1980s and beyond, even though at present it is not as widely accepted in the United States as in some other countries.

Viewdata has two forms. In the simpler, sometimes called teletext, data from computer-based storage can be retrieved on command using a television receiver which has been provided with enhanced electronics. Input is not possible, except for commands which select the specific data to be displayed. Applications for which teletext is suitable include the display of airline, train, or bus schedules; the listing of items (homes, automobiles, etc.) for sale; and telephone directory listings.

The other form of viewdata is interactive and allows both the selection of items to be displayed and the input of data based on what is shown on the screen. Applications for interactive viewdata include booking an airline, train, or bus reservation; buying a ticket to the theatre; and shopping at home for groceries, clothing, etc., advertised on the TV. All these examples would typically also involve either automatic charging to a credit card or electronic funds transfer to effect payment.

Viewdata in either form involves three components:

- A television receiver with expanded electronics and an enlarged remote-control keypad for selection and data input

- One or more computer systems to provide the database of displays and to process interactive input if allowed

- A delivery mechanism over which the data are transmitted

In some cases a display terminal is used instead of a television receiver, but in general the term "viewdata" implies the use of TV as the receiving and sending terminal device.

The question of which delivery mechanism(s) will be used for viewdata is a topic of hot discussion, especially in the United States. There are three choices: broadcast TV, cable TV, and the telephone network. The advantage of using broadcast TV is that television sets already use that method and little additional cost is required to add viewdata capability. The limitation of this mode is that it can support only displays, not interactive applications. Cable connections can support both teletext and interactive applications, but not all TV sets are connected to cable at present—and in the United States there are a variety of cable systems, not a common network. Use of the telephone network requires additional complexity in the TV electronics, so that the receiver alternately can be connected to a telephone line or to cable or be allowed to receive broadcasts. The advantage of this method is that almost every home and office is connected to the telephone network (at least in the United States, and increasingly in other countries) and both interactive and display forms of viewdata are feasible. Note, however, that the bandwidth (transmission capacity) of most links in the public telephone networks is much lower than TV broadcast or cable, which effectively limits viewdata to fixed-frame video. Broadcasting motion, as in TV programming, requires a higher bandwidth. Viewdata today is limited to fixed-frame, meaning that each frame is transmitted and displayed separately, since frames are essentially data presented in image form. However, in the future viewdata is likely to evolve to the use of motion, especially in applications directed to the home rather than to data input/output situations.

Viewdata is in the very early stages but already shows promise. For example, in the United Kingdom one application which has been studied is the use of viewdata for order entry in breweries. Today each pub owner periodically telephones orders to the brewery's distribution center. The distribution center may simply record the order, without being able to tell the pub owner when the shipment will be made. Since most pubs have television sets installed, adding viewdata (perhaps at the brewery's expense) would allow order entry and acknowledgment via viewdata. In this case the pub owner would dial into the brewery's computer system, rather than to the distribution center, and call up screens of product-description data. When a desired product was displayed, the amount needed could be entered using the TV's remote-control keypad, and a required date could optionally be keyed in. The computer could then respond immediately, on the screen, with an expected delivery date or back-order notice. An example of this type of viewdata dialogue, with the input data underlined, is shown in Figure 9-7.

DESIGNING INTERACTIVE INTERFACES

Selecting from the devices and methods described in the preceding section, user interfaces must be designed for the specific application being studied. This section describes how to select devices and methods for interactive interfaces; the following section briefly discusses batch-mode interfaces.

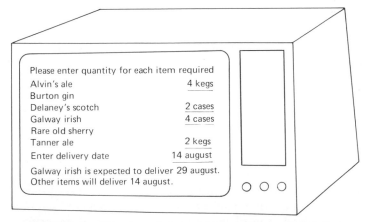

Figure 9-7 Viewdata dialogue (user input underlined).

Selecting interactive terminal device(s) and method(s) may be easy or complex, depending on the situation. In too many cases, system designers do not consider all the possibilities and decide to use video-display terminals because those are the most commonly used devices. However, a number of factors ought to be considered before selecting the type(s) of terminals to be used. The relevant factors are presented in the form of a series of questions, the answers to which will indicate which type(s) of devices and which method(s) are best suited to the application. These questions must be answered individually for each group of users if the groups have different requirements, backgrounds, and/or working environments. The conclusion may be that different devices and/or different usage methods are appropriate for different classes of users.

1. *Is there a special-purpose terminal device available which will meet the users' needs?*

Although not all applications require special-purpose devices, if any are available they are worth serious consideration. Special-purpose devices are designed to simplify the user's interface with the information system, which is, of course, one of the major goals in system design. If a terminal manufacturer has already performed the analysis necessary to define a specific class of users' needs and then implemented a terminal to meet those needs, the individual designer need not repeat the entire process of terminal-interface design. Of course any evaluation of special-purpose terminals is somewhat like the evaluation of preprogrammed application logic. Although special-purpose terminals are designed for a specific application, or set of closely related applications, they are by definition somewhat generalized so that they can be used by a number of organizations with similar application requirements. Any specific organization may find that it has needs which are not met exactly; it is then necessary to weigh the advantages of acquiring devices designed for that environment against the possible disadvantages of changing methods somewhat to make use of the devices as designed. In some cases this trade-off will result in the decision to use general-purpose terminals instead, so that the interface can be tailored more specifically to local needs. In other cases the result will be a decision to modify methods and procedures as necessary to make use of the available terminal devices.

Another possibility, which was mentioned earlier, is to have terminals designed and built specifically for the application being studied. This is practical only if a very large number of terminals is required and the cost can be justified in terms of improved efficiency. In most cases, low volume—which would result in a high cost of manufacture per terminal—makes special-purpose, customized devices impractical.

Most new terminal devices are microprocessor-based and can be modified by changing the program embedded in the device. If an existing terminal is close to what is needed, it may be useful to evaluate the possibility of changing the firmware to more exactly match local needs. The terminal being considered may be either a special-purpose or a general-purpose device; in either case the vendor may be willing to consider firmware change if the volume is sufficient to justify the expense.

If a special-purpose terminal is chosen, its design and features will usually make the design of user interfaces relatively simple. Because devices in this class represent the manufacturer's accumulated experience with similar applications, the design of user interfaces for a specific application is typically easier than when a general-purpose device is used.

2. *What is the working environment of the users?*

Factors such as the noise level which exists and/or which can be tolerated must be studied. (These environmental aspects are often called *ergonomic factors*.) If the environment is quiet and noise is a potential problem, quiet terminals such as displays are most appropriate and the use of hard-copy devices must be minimized. If printed output is essential, the printer(s) ought to be placed in other location(s) or enclosed in noise-deadening covers. A high existing noise level may make voice input and/or output impractical because of the potential interference. In contrast, if the noise level is low, voice input may be a potentially disruptive element, and the installation of sound-deadening facilities such as fabric-covered partitions and carpeting may be necessary.

If voice input or output is considered, the aspect of security and privacy must be taken into account. Voice input of sensitive data may lead to disclosure if there are people in the area who might eavesdrop; voice output can similarly disclose sensitive information. If voice output is provided via telephone, security or privacy protection can be concentrated on the transmission mechanisms, as there is little danger of disclosure from overhearing the actual output. However, if the voice output is from any type of speaker, eavesdropping can be a problem. This concern also applies to output provided by the system to validate voice input (as described earlier); this may lead to the disclosure of sensitive data if not properly controlled.

If the environment is relatively "hostile" to terminal equipment, as in a factory, warehouse, loading dock, or similar location, care must be taken to select only terminal devices which can operate successfully in those conditions. Terminals designed to operate in environments of this type are *ruggedized*, meaning that they are specially packaged in different enclosures from normal terminals. This packaging allows them to operate where dust, humidity, heat and/or cold, or vibration would disable most terminal (or computer) equipment. If there are no special-purpose devices available for the specific application or those which are available are unsuitable, it will be necessary to either (1) obtain specially built or modified terminals which can operate in the environment or (2) redesign the work flow so that users can access the terminals in a less hostile environment. This problem is less acute today than a few years ago, as equipment now is generally more resiliant and terminals can typically operate with wider environmental tolerances than was the case in the past. However, there are definitely limits to the extremes which can be tolerated, and if those limits are exceeded, other alternatives must be found.

Other aspects of the environment must be considered, some of which were discussed in the context of data collection (see Chapter 5). If the users will be serving customers or clients and will be required to carry on a dialogue with those people while using a terminal device, interface procedures must be very simple and easy to use, as distractions will occur frequently. In those environments, which may also require rapid response, keying ought to be kept to a minimum. Special-purpose devices or the use of techniques such as badge entry, card entry, or document entry may simplify the user interfaces. If a general-purpose VDT is chosen, forms-mode interfaces are most likely to meet the need for high speed, simplicity, and a high degree of accuracy.

Although space is often ignored at this point in the system design, it may become a problem later if not evaluated now. If the environment in which the users work is crowded already, the introduction of terminal devices—often accompanied by cluster controllers and sometimes by minicomputers—will make the situation more critical. The problem is especially acute when terminals, controllers, printers, etc., are installed in an office. Many offices are already crowded, and it may not be easy to fit the new equipment into the available space. If this situation exists, either terminal devices which can fit into the current area must be chosen or some way must be found to move the users into more spacious quarters. Ignoring this problem or believing that some way will be found to solve it after the terminals are installed is asking for serious trouble.

Special attention must be given to the environment of users who will use terminals continuously, as they may develop symptoms of physical and psychological stress if the interface design and physical environment are not well-planned. For example, users in a word-processing center who access video-display devices all day may complain of eyestrain if the lighting produces glare on the screens. Well-designed ambient and/or individual work station lighting plus the selection of display-screen colors appropriate to the environment can minimize this problem. (Many VDTs offer a choice of green or black background, and some offer other colors, plus the choice of either white characters displayed on a colored background or colored characters on a white background.)

Other potential problems include neck and back strain caused by poor seating arrangements and/or the wrong height for the terminal work surface. Since various users have different lengths of arms and legs, the ability to lower and raise work surfaces (as well as chairs) is important. These problems are most often felt by dedicated users but may arise in situations where the user accesses the device only intermittently. Occasional users, however, are less prone to stress of this type, since they usually have the freedom to get up and move around at any time that terminal use threatens to become uncomfortable. Users who do not have that freedom are much more likely to develop problems; however, these can be foreseen and minimized with careful planning.

It is impossible within the space limitations of this volume to provide a complete, detailed survey of all the aspects of the environment which affect user-interface design. The important point is that it is essential to thoroughly study the users' environment in each case and determine in detail the impact of the surroundings on the devices and interface methods which will best serve the users' needs.

3. *What are the response-speed requirements?*

Fast response, especially when the users will be interacting (in person or by telephone) with customers or clients, requires simple interface procedures. If special-purpose terminals exist for the application, they will probably provide the quickest response, because of their tailoring for ease of use. Voice input and/or output may also be appropriate in this case, since most people can speak more rapidly than they can enter data either by keyboard or by special-purpose devices—unless the input is very simple, such as selecting one choice from a menu displayed on the screen. As discussed, voice equipment may be inappropriate, based on other aspects of the users' environment and requirements, even if response speed makes it an attractive choice. If neither special-purpose devices nor voice is practical, VDT devices are typically chosen when fast response is needed.

Applications which do not require such fast response, in contrast, allow much more flexibility as to which terminal devices are chosen. Essentially any type of device may be selected, based on the other criteria listed.

4. *What level of accuracy in the user interfaces is required?*

Too often the automatic response to this question, if it is even asked, is "absolute," or "100 percent." Those responses, although understandable, are unrealistic. Certain types of data must be as accurate as it is technically and humanly feasible to make them; others can be inaccurate without causing serious problems. In a payroll application, for example, an inaccurate paycheck amount will result either in a loss to the employer or in difficulties in the relationship with the employee, depending on who is favored by the error. On the other hand, a slightly misspelled employee name may be unfortunate, but the consequences are less severe. If payroll checks are mailed to employees, an inaccurate address may result in a lost check, but if the checks are handed out at the workplace an erroneous address is less serious.

To amplify this point, data entry for data-processing systems and the entry of text into word-processing systems are technically very similar; in each case a person keys in material using a terminal device. However, data-processing input often consists mainly of numerical fields which will be used in calculations and which must have a high level of accuracy. Many data-entry systems therefore provide for key verification of the data; that is, another terminal user rekeys all (or the important parts) of the same data and the new input is compared to the old. If there is any difference, the data-entry device signals a possible error, and a correction can be made if the error was in the original entry.

Text input, in contrast, often consists of memos, letters, reports, and similar documents, in which the majority of the material is textual rather than numerical. Although accuracy is desirable, errors such as misspelled words are not significant enough to justify the added expense of completely rekeying the text. If numerical data are included, the accuracy is typically verified by reading the data as displayed on the terminal screen. As systems move to the handling of more textual material, accuracy concerns may shift and the ability to distinguish between elements which must be accurate and those which need not be will become more important than it has been in many data-processing systems.

After a decision is made about the level of accuracy required, this is one of the factors used to select the terminal type(s). If very high accuracy is required, then data entry on a VDT, using forms mode with immediate validation of data items entered, is appropriate. In some cases, a separate verification step by a second person may be necessary, although obviously this must be avoided whenever possible because of the added cost and delay. Hard-copy terminals typically do not provide the same level of accuracy in data entry, both because they do not provide an equivalent of the VDT's forms mode and because it is generally harder for the user to notice errors in printed material than on a display.

Special-purpose terminals, which provide positive ways to input certain types of data (via badge reader, for example) and which may allow a single function key to be depressed to replace keying several characters, can contribute to accuracy if well designed. Voice input, in contrast, is still less accurate in most applications than other types of input. However, as the earlier discussion of General Electric's plans for the use of voice-recognition devices in factories illustrates, in specific circumstances the use of voice may be at least as accurate as other methods in the same environment.

Two ground rules must be kept in mind when designing user interfaces in which accuracy is very important. First, minimize the amount of key entry required, since fewer key strokes will statistically result in fewer errors. (Of course, minimizing key strokes is a very good rule for other reasons but is especially important to raise the level of accuracy.) Second, design extensive editing and validation routines to ensure that the data entered are correct, within the practical limits of editing routines. Limit and range checks can determine that amounts are reasonable; check digits can help to ensure that an account or identifying number is not only a valid number but also the correct one for the specific transaction. Some types of data, such as descriptive fields, cannot be checked in most cases, but errors in those fields are typically less disruptive than in other fields.

5. *Are there requirements for printed records because of legal and/or auditability needs?*

The data-collection process in Chapter 5 ought to have defined any requirements for hard-copy records. In some applications, for example, it is necessary to retain a printed copy of all transactions for some defined period of time. (Some applications do not involve a mandatory hard-copy record, and it is acceptable to keep a computerized copy, such as a disk or tape file.) If requirements of this type exist, the system design will have to provide for the necessary output.

Such requirements may make it appropriate to configure hard-copy printers at the user locations, so that input can be captured or output provided during interaction with the user. In other cases, it will be possible to provide the necessary printout on a periodic basis from the computer(s) which process the users' input. If feasible, the latter approach is preferable, since it avoids the need for printer devices at the users' work stations. Printers cause noise, they add to the cost of the users' work stations, and they also create printed output at dispersed locations where it may be difficult to manage. Centralized printing, or at least printing at a minimum number of locations, is much better if the system design can be arranged to allow this.

6. *What kind of background do the prospective users have?*

In an office or in service industries such as banking, finance, insurance, and so on, many people have keying skills or the necessary background and attitude which make the teaching of these skills possible. In this situation, the use of video-display and/or hard-copy plus keyboard terminals is probably appropriate (if there are no special-purpose devices which are suitable). Note, however, that in many office applications the terminals will be used by managers, professionals, and executives, who may have very negative reactions to being asked to key in data. (This is not universally true; some executives adapt to the use of VDT terminals very readily.)

In environments such as factories, distribution warehouses, and service stations, it is unlikely that prospective users will have keying skills. Like many managers and professionals, those people will typically be very negative about any attempt to make them learn to type. Every effort must therefore be made to define interfaces which require minimal keying. The use of badge readers, document readers, and/or card readers for input ought to be investigated. If voice-input devices are available which might be suitable, those also ought to be considered. Other techniques, such as touch-sensitive display screens with tutorial, menu-driven interfaces, may be well-suited to these environments. With careful design, some set of these techniques can provide interfaces involving minimal, or in some cases no, keying.

The educational level of the users must also be taken into account, both in determining how easy to use interfaces must be, and in defining the exact vocabulary to be used for input and output. (This topic was covered in Chapter 5, and the data-collection process ought to have provided a good profile of the users' educational and cultural backgrounds.) In any case, the use of data-processing jargon must be avoided. Simple vocabularies are often important, as the users may not have a high level of education. Customization to meet local needs may also be important. For example, people who live in certain locations in the United States use different terms from those in other locations; this may be an element of the user-interface design. Different English-speaking countries do not use identical vocabularies; the same word may have a different meaning in England, the United States, and Australia. If the users have their own specialized vocabulary, as in the case of the medical, legal, and engineering professions, the correct terms must be used when designing interfaces. Sensitivity to the appropriate vocabulary and "tone" of the user interfaces will help to ensure that they are easy to use and acceptable to the users.

7. *What level of turnover in the user work force is expected?*

If the user work force is relatively stable, then fairly complex interfaces may be practical, within the constraints established as a result of answering the earlier questions. A stable work force with keying skills will find the use of video-display terminals very comfortable. A work force with rapid turnover probably will not, especially if keying experience is minimal. Rapid turnover makes it essential to establish the simplest practical user interfaces, making use of special-purpose terminals whenever possible.

The expectation of rapid turnover also emphasizes the need for a training mode. This mode allows an individual terminal to be taken out of production mode and attached to special training routines, so that a person can practice using the terminal without affecting production databases and without interfering with other, i.e., production-mode, users. Training mode may include both educational text, in effect a user's manual available via the terminal, and practice transactions with tutorial explanations and expanded error messages. Although training mode is especially important if a high rate of turnover is expected, it is very useful in every interactive system. A well-designed training mode minimizes the need for one-on-one educational sessions with new users and is also helpful when experienced users occasionally encounter unusual transactions. The system design must make it possible, on user request, to switch from production mode into training mode and back.

To summarize the process of selecting the appropriate terminal device(s) and method(s) for interactive interfaces, Figure 9-8 shows how the various factors which are covered in the preceding questions relate to the design of interfaces. This figure can be used during the design process as a guide to which choices will probably be best suited to each type of environment and set of user requirements. No column for viewdata is provided in this figure because there are no applications yet for which this is the preferred mode—except when the user is in the home. More experience with viewdata for business applications will be needed before firm guidelines can be provided.

To reiterate the importance of prototype interfaces, if these have not been implemented at some earlier point in the analysis and design process, they ought to be now. If terminals of the type which will probably be used are already installed, they can be used to provide mock-ups of the major interfaces, so that the users can evaluate how they will work. If new types of terminals will probably be acquired, one or two devices of that class (although not necessarily of exactly the model which will be chosen) ought to be obtained—on a lease arrangement if possible. These can then be used to create prototype interfaces for the users to evaluate. Even if prototype interfaces were set up earlier in the system study, any change in approach which occurs during the analysis and design process ought to be reflected in a revision of the prototype. Otherwise, decisions made on the basis of an earlier prototype may prove invalid, and the final design of the user interfaces may not be satisfactory.

| Selection factor | Special-purpose | VDT | | | Voice | Printer |
		Forms mode	Menu mode	Natural language		
			Terminal types and methods			
Special-purpose device available	Y					
Hostile environment	P	P	P		C	
Security or privacy protection	P	P	P		C	
User interaction with customers or clients	Y	P		N		
Fast-response requirements	Y	P	P	N	P	
High accuracy	Y	P	P	N	C	
Printed records needed						Y
Users do not have keying skills	Y		P	N	P	
High user turnover	Y		P	N	P	

Y = Indicated
P = Possible choice
C = Possible choice, but care required
N = Definitely not indicated

Figure 9-8 Designing interactive interfaces.

DESIGNING BATCH-MODE INTERFACES

It is unnecessary, during strategic-level system design, to define batch-mode interfaces in the same detail as interactive interfaces. The batch-mode interfaces consist of report formats and the method by which those reports are requested. In many systems, reports are requested procedurally, through an agreement between the users and the IRM operations staff that each report will be produced on a specific schedule; in that case no design effort is involved. In other cases, the reports may be requested by interactive or job-control-language input supplied by the users when the report is needed; in that case the specific method of making the request must be defined.

At this point in the system design, it is only necessary to define which types of reports will be needed, their frequency and production schedule (if any), and generally the data elements which will make up the content of the report. Much of this information will have been accumulated during the process described in Chapter 5. Now that the system analysis and design have progressed much further, it is appropriate to reevaluate those output definitions and modify them if necessary, based on the additional understanding of user requirements. New requirements may also have come to light during the system study; if so, they must be documented.

Among the decisions to be made at this time is how output will be prepared. In most cases high-speed printers (either impact or nonimpact) are used to produce high-volume output reports. However, if reports are required only for legal or auditability purposes it may be unnecessary to print the data. Output on microfilm is an alternative which may be very attractive, because less space is required to maintain records on microfilm than on paper. Even output which is required for more immediate use but which is very voluminous may be well suited to microfilm. For example, source listings of large software systems can be stored on microfilm, with microfiche readers made available to people who need to consult the listings. This method does not work well if many people need to use the output frequently, but with a low frequency of access it is very effective.

Another point to be considered is the need to minimize batch output by reevaluating requirements and redefining them if possible. One of the well-known attributes of computer systems is that once the production of a batch report begins, it is very difficult to discontinue. Even though no one may be using the report effectively, the recipients will often vigorously oppose an attempt to stop printing it. A good rule, therefore, is initially to prepare the absolute minimum number of reports. If other reports are necessary, they can be added later. However, a complete analysis of the likely user data needs is essential, as reports which require data elements not planned for in the system will be very difficult to implement later. A good plan is to define all probable data needs, but initially implement only the batch reports which have a well-defined purpose.

The best alternative to batch reports is a set of ad hoc query and reporting capabilities. If users understand that it will be relatively quick and easy to obtain data which are not provided in routine reports, they will usually agree to minimize the number of reports printed. If, on the other hand (as is typical in many installations), users know that it is difficult, costly, and time-consuming to obtain data from the information system, they will usually insist on voluminous reports containing all possible data elements. Recognizing this syndrome is the first step in avoiding it, and the second step, as noted, is to provide ad hoc capabilities. These ought to be an essential part of every new computer-based information system.

The final point to be considered at this stage in system design is the need for flexibility in batch reports. The content, format, and schedule on which reports are prepared must all be designed as parameters which can be changed when requirements change. This is less important when ad hoc methods are used for many data requirements than when all needs are met via batch reporting; however, flexibility is still essential. Requirements defined prior to system implementation are very likely to change, perhaps even before implementation is complete. It cannot be repeated too often that flexibility is an important element of information systems in the 1980s and beyond, and this applies to batch-report methods as well as to the interactive aspects of the system.

SUMMARY

The interface definitions which are the result of this step in the system-design process must be formally reviewed with the users and their management, and formal agreement must be reached that these interfaces will meet the users' needs. If only a small number of different interfaces are required, as for example in a reservation system, then a single review meeting may be held, at which the interface specifications are presented and the prototype is demonstrated. If the specifications and prototype are acceptable, a formal agreement can be signed immediately by the user management and the system designers. If modifications are agreed on, the review can be repeated at a later date, if necessary, and an agreement signed then. In a complex system which includes a wide variety of different interfaces, several reviews may be appropriate, each involving one group of users or some specific types of interfaces. Agreements can be reached separately at each of these reviews, or a single agreement can be prepared when all reviews have been completed successfully.

In all cases, the parties to the agreement must explicitly agree that changes may still be necessary. From the users' side, changes may be needed because parameters affecting the requirements change. From the system designers' side, changes may be necessary because more detailed system design and/or implementation efforts make it clear that it is infeasible to provide exactly the interfaces agreed on—or it may become clear that improvements can be made to the interfaces. This flexibility to change, for a variety of reasons, must be written into the agreement. In addition, the result of change, which is often delayed schedules, must be spelled out. Time taken to clarify in the agreements how needed changes will be handled is an important part of building a good relationship between users and designers.

10

CHOOSING THE
SYSTEM STRUCTURE

Defining whether, or how, to allocate system functions across multiple information-processing elements is one of the most fundamental design decisions. The predominant form for information systems in the 1980s and beyond will be what is today called *distributed data processing* (*DDP*). However, there are some applications which are better suited to a centralized-processing structure and some which are best implemented as multiple freestanding, decentralized systems. This chapter discusses how to decide which approach to use in the new system. If a distributed structure is chosen, the earlier volume, *The Distributed System Environment,** will provide additional detail concerning different system structures—hierarchical, horizontal, hybrid—and how each type can best be used.

Although this chapter discusses the choice of a system structure as a relatively independent topic, in practice it is closely related to the database-design decisions addressed in Chapter 11, and to the network-design process described in Chapter 12. System-structure decisions are also related to the techniques necessary to design for integrity and flexibility, which are discussed in Chapter 13. Chapters 10 through 13 must therefore be viewed in context; each addresses one aspect of the fundamental structural design choices which must be made in a complex system. In practice, those decisions will be made during an iterative process in which the effect of decisions in one area (e.g., database design) are evaluated for impact on the other areas and often modified as a result of the evaluation.

It is, of course, important to remember that the system-structure decisions must be made within the framework of (1) the existing system structure(s) of related applications and (2) the organization's policy (if any) concerning the distribution of functions.

* McGraw-Hill, New York, 1981.

If the new application is closely related to one or more existing applications, then the required flow of data to and/or from those applications may make the best choice to collocate the new functions with the related functions. If the existing functions are centralized, there is a strong probability that the new functions also ought to be centralized. If the existing functions are distributed, then the new functions will probably also be distributed and follow the same method of partitioning as the existing functions.

Many organizations have adopted guidelines which identify preferred information-system structures and how to decide which one to choose in specific situations (see the discussion of long-range plans in Chapter 2). The organization's guidelines may have been formulated by considering questions similar to those in this chapter and answering them in terms specific to that organization. In some cases, organizational guidelines will provide a ready-made decision about the system structure; in other cases, additional factors, such as those discussed here, must also be considered. If the organization does not currently have this type of guidelines, experience gained during the system-design process can be used to help establish ground rules so that future choices can more easily be made.

There are many factors involved in deciding whether to centralize, decentralize, or distribute application-related functions, and the decision must be made by evaluating those factors and their interrelationships. The relevant factors are presented here in the form of a series of questions; the answers to these questions, as they relate to the system being studied, will indicate the most appropriate system structure.

CAPACITY

1. *Are the capacity requirements of the application too great to be implemented in a single computer system?*

At this stage in system design the application's capacity requirements ought to be reasonably well defined. The capacity needed is determined by the total number of users to be served, the number of users who will simultaneously access the system, the volume of transactions to be handled, the response-time requirements, the size and complexity of the database, and the complexity of the processing functions (and of course the amount of overhead required for system functions such as the operating system, integrity protection, and so on). Although the definition of capacity requirements is typically still incomplete at this point in system design, certainly the magnitude of the requirements must be understood.

Expected growth must be included in evaluating capacity needs. Allowing at least 10 percent unused capacity for growth is always good practice, and a great many systems require a higher percentage than that. The data collected in Chapter 5 concerning expected, probable, or possible future changes will be very helpful in deciding how much growth is likely and therefore what level of capacity increases must be handled. Expected growth is especially important if a centralized-system structure is being considered, because the rapid expansion of a centralized system is typically more difficult than comparable expansion in a distributed or decentralized system. Of course some large-scale computers are structured so that additional processing modules, memory modules, input/output modules, and peripherals can be added without seriously affecting modules already installed. However, for some other computers it is necessary to replace an existing processor or other module when a more powerful model is installed. In these cases, disruption is very likely. If capacity is expanded by duplicating the existing centralized-system hardware to increase capacity, then in effect the system is converted to either a decentralized or a distributed structure.

Allowing a cushion of unused capacity is most important in systems which provide ad hoc query and reporting facilities. These tend to attract new users very rapidly, provided the facilities are easy to use, and this kind of growth is very difficult to predict accurately. Another situation which can cause rapid and unpredictable growth is the implementation of office-related functions such as electronic mail. These facilities, like ad hoc queries, tend to be habit-forming once users have discovered their advantages.

It is possible that the required capacity, including adequate provision for growth, can be supplied on an information processor already in use by the organization. That is a very attractive possibility and may lead to a de facto choice of system structure. However, even then the remaining questions ought to be analyzed to ensure that other factors support the decision to centralize functions.

Capacity requirements and/or the need for rapid expansion may dictate the need to explore distributed or decentralized alternatives because the requirements—present and/or future—cannot be met on one system. Of course one solution which is often chosen is to use multiple computer systems, collocated in a single center, and partition the work load across them. In many cases, this is really a decision to select a distributed-system structure, even though the "communications links" between the systems may be people carrying reels of magnetic tape for data exchange. More often today, configurations of this type are connected via high-speed bus, forming one type of local-area network (LAN) (LANs are discussed in Chapter 12). In other cases the application can, on closer examination, be partitioned to fit on two or more decentralized computer systems. However, as the growing complexity of business, government, and society in general causes an increased volume of data flow among applications, fewer and fewer decentralized systems will prove viable.

If the evaluation of capacity requirements does not specifically indicate that a distributed structure is required, the other questions in this chapter must be evaluated.

FLEXIBILITY

2. *Is a high degree of flexibility for change in the functions, methods, or data structures of the application required?*

The information about future change, gathered in Chapter 5, is useful not only in evaluating the need for capacity growth (question 1) but also in determining how likely it is that other aspects of the system will require modification. The greater the degree of expected change, the less likely that a centralized-system structure will be advantageous. A centralized system typically encompasses a large number of application functions, possibly unrelated except for the fact that they are executed on the same computer equipment. A centralized system of this type can be extremely complex, and one of the basic rules of information systems (computer-based or manual) is that the more complex the system, the more difficult change becomes. Although it is theoretically possible to change any system, in practice change may be very difficult, and it will therefore be impossible to change quickly. Information systems which serve main-line operational functions must be able to change as rapidly as the organization's environment changes, so inflexibility can be a major handicap.

Distributed and decentralized systems, in contrast, are by definition modular. When well designed, these structures can be changed much more rapidly than is typically the case with centralized structures. The need for rapid and/or extensive change is therefore an indication that a decentralized or distributed system will be a better choice than a centralized system.

AVAILABILITY

3. *Are there requirements for a high level of availability and/or a high degree of resilience to failures?*

The requirements for availability were defined during data collection (see Chapter 5) and evaluated during data analysis (see Chapters 7 and 8), which may have identified clusters of functions that require high availability. In many new systems, high availability and great resilience are essential because the organization will be heavily dependent on the continued operation of the computer-based system.

Availability and resilience are closely related but not identical. Availability is technically defined as the percentage of total scheduled time (which may be less than or equal to 24 hours a day, 7 days a week) that the system is usable and can carry out the required work. Resilience (or survivability) is a less formalized way to express how likely the system is to be able to survive various types of errors or failures and continue to perform useful work. Both are important in on-line systems.

Ideally, availability and resilience ought to be defined in terms of each individual user's view of the system; that is, the probability that a user can obtain the needed functions and data from the system whenever appropriate. Realistically, however, the best that can be done is to consider each unique group of users and evaluate the needed level of availability and resilience from their viewpoint. The requirements of each user group ought to be well defined as a result of the data analysis performed earlier.

There are two ways to improve availability and resilience: one is to provide two or more of a component (hardware, software, data, or facility) which is likely to fail; the other is to separate functions and components to reduce vulnerability to single-point failures. If high availability is required in a centralized system, the approach usually chosen is to duplicate all the hardware—and sometimes the software and data as well. Elaborate, and costly, precautions against unauthorized entry or actions are also typically taken. While it is possible to guard effectively against certain types of problems, such as hardware failures, the cost of this protection is often high.

Centralized systems are, by their nature, vulnerable to single-point failures which affect the location where they operate. Of course there are many ways to minimize this problem; for example, vulnerability to electrical power failures can be avoided by installing battery backup and generating capability on-site. However, natural disasters such as fire, flood, etc., and unnatural disasters such as sabotage are more difficult to guard against effectively, and the precautions taken may be very expensive.

In a distributed or decentralized system, it may be possible to achieve the level of availability needed by selective—rather than total—duplication of components. Only the hardware, software, and data which are essential to continued operation can be backed up, while less expensive protection methods are used for other elements. Distributed and decentralized systems are typically dispersed over multiple sites and are therefore less vulnerable to single-point failures than are centralized systems.

The need for a high level of availability and resilience is therefore an indication that either a distributed or a decentralized structure will be the best design choice. Chapter 13 discusses in more detail how to design for availability and resilience.

PROCESSING MODE

4. Which mode of processing will predominate, interactive or batch? (In some cases real-time processing, such as process control in a refinery, must also be considered.)

The more interactive the system, the greater the likelihood that the distribution of at least some functions will be advantageous. Batch-processing systems are not good candidates for functional distribution, as several of the major advantages of distribution—improved response speed, higher availability, ability to tailor for local needs, ability to distribute managerial control, and reduced cost of communications—apply poorly or not at all to batch systems. Systems which are predominantly interactive and/or which handle real-time input and output are good candidates for a distributed-system structure. The other factors discussed in this chapter will then determine if that is the best choice.

CLUSTERING

5. Are there clusters of shared functions, each of which serves a separate group of users?

Data analysis (see Chapters 7 and 8) may have identified patterns of functions which can be quite easily translated to a distributed-system structure. If a group of users has the same functional requirements, and especially if the group is geographically separated from other groups, the installation of an information processor to provide the necessary functions is often attractive. If the group's data-access needs also partition cleanly from those of other groups, a distributed-database partition (or replicated segment) can be attached to that processor. Matching the system structure to the patterns of user requirements, whenever possible, is a basic design technique in complex systems.

If these patterns exist but each group of users has completely independent requirements, with little or no need for data exchange with other groups, then a decentralized structure is probably best.

CUSTOMIZATION

6. *Are there a number of common user requirements, but also users or user groups who need somewhat different functions, methods, interfaces, and/or reports than the other groups?*

Even though a large number of users may share the same *general* set of requirements, some users or groups of users may have different *specific* requirements. For example, a bank may have multiple branches, all of which provide the same set of services to their customers. However, because of its geographical location, one branch may have a much higher than average proportion of savings accounts to checking accounts. The manager of that branch might need somewhat different reports than do other branch managers, to support the analysis of customer trends for use in defining the correct marketing programs for that branch. As another example, several factories might build the same products but local conditions might call for different procurement procedures, different hiring methods, and different pay scales.

Designers of computer-based information systems too often attempt to smooth out differences so that a single processing method can be used. Of course, if the differences are significant enough, such as the use of different labor contracts or different pay scales, they cannot be ignored. However, more subtle differences—such as a manager's need for customized reports or the ability to tailor input formats to users' preferences—are often considered frivolous and viewed as the cause of unnecessary added cost. This is particularly true when all users are served by a single centralized-processing system.

The result of this habit of forcing users, methods, and procedures into a single mold for the sake of computer-system efficiency is often decreased human efficiency. To some degree this mode of thinking reflects the very different environment of 15 or 20 years ago. At that time, computer equipment was extremely expensive, and every unnecessary processor cycle could add significantly to total cost. Wages and salaries, in contrast, were lower than today, both in an absolute sense (in constant dollars) and relative to the cost of computer equipment. Much emphasis was therefore placed on conserving computer resources, even if the result was inconvenience to the users.

Today's information-system environment is dramatically different from that of the past. The cost of computers and related hardware has dropped precipitously, both relatively and in constant-dollar terms, while wages and salaries have escalated sharply. In addition, many more people are using computers each year, and often using them as essential tools in carrying out their duties. False economies in processing equipment, to gain computer efficiency, are therefore out of touch with today's reality.

System analysts and designers must make a special effort to be sensitive to the need for customization throughout the data-collection, data-analysis, and system-design processes. Users must be encouraged to state their needs, with the understanding that perhaps not all of them can be met but that a genuine effort will be made to meet as many as possible. The close cooperation between designers and prospective users, which is repeatedly stressed in this volume, will make it possible to handle customization requirements appropriately.

Whenever customization is a stated requirement, and especially if the needs for tailoring generally match the patterns of functional requirements discussed earlier (see question 5), a distributed-system structure is probably the best choice. The ideal situation for cutomization, although unfortunately one which does not always occur in the real world, is shown in Figure 10-1.

This example shows three groups of users, in three geographically separate locations of the same organization. The users all share the same functional requirements for the information system, but local variations indicate the need for some functions, some user interfaces, and some report content and formats to be different at each location. In this situa-

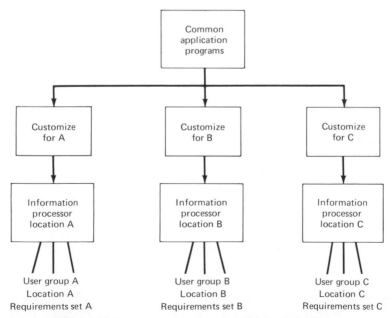

Figure 10-1   Customization in a distributed system.

tion, the central development of common programs is appropriate. These common programs can then be customized to meet the specific needs of the users at each location, with the modified programs sent to the processors at the three locations. Changes to the common programs must of course be reflected at all locations and evaluated in each case to determine if there is a need for further customization.

It is, of course, perfectly possible to provide this same type of customization even if all three user groups are served by a single central computing system. However, this is seldom done in practice, and customized functions may be more difficult to maintain in a centralized system than in a distributed approach. (Customization in a centralized system results in increased complexity, which usually limits the flexibility to change; see question 2.) On the other hand, simply selecting a distributed-system structure does not guarantee that the needed customization will be done; constant attention to user needs is necessary to ensure this.

In the example system, customization could also be done by the local user groups, if the management philosophy and user skills allow this. Some organizations prefer not to allow local programming, although with the advent of personal computers this restriction is becoming harder and harder to enforce. Some user groups also have no desire to learn any form of programming, even to meet their needs for customization. Others, however, are very willing to do so. Perhaps the best compromise is to create common programs with parameter-driven interfaces, both for input and for output and reports, including content and format. Local groups can then change the parameters as desired, and the central IRM staff need not be directly involved in customization activities. The wider the range of variation supported in this way, the more likely that the majority of customization needs can be met.

SECURITY AND/OR PRIVACY PROTECTION

7. *Will the system handle data and/or functions which require tight protection for security or privacy reasons?*

The analysis of security- and privacy-protection requirements (see Chapters 7 and 8) may have indicated ways to partition sensitive data or functions from others. In that case, the segmentation may allow a distributed structure to be designed around the partitions—so that protection can be concentrated on the sensitive areas. Protection inevitably adds to costs, and it is therefore important to protect only where necessary.

The need for tight protection may suggest that a centralized system is best, or if a distributed structure is chosen, it may remain a locally distributed one; i.e., collocated computers in a computer-center environment. The physical protection provided by a restricted-access site, as in almost all computer centers today, makes it easier to handle sensitive data or functions. If only certain functions and/or data elements are sensitive, then locating those at a central site may be wise; other parts of the system can be distributed or not, based on other criteria.

Although locating protected data and functions centrally may be the easiest approach, it is by no means a complete solution to the problem. In many cases remote access must be provided to those data and/or functions, yet they must not be compromised. Also, as increasingly larger numbers of functions and amounts of data are moved onto computer-based systems, protection through centralization will become progressively less viable. Organizations must therefore become accustomed to planning for distribution while still providing an acceptable level of protection. This topic is discussed again in Chapter 13.

GEOGRAPHIC CONSIDERATIONS

8. *Are the users widely dispersed geographically?*

If the answer is yes, chances are that a distributed system will lower the cost of communications, and it may also improve the speed of response to terminal users. (In some cases decentralized computers may provide the same benefits.) "Widely dispersed" does not necessarily mean that users are separated by thousands, or even hundreds, of kilometers. Users distributed around the cities and towns within one state in the United States may present a good pattern for distribution, especially if the state has higher-than-average rates for communications (states vary widely in their intrastate rate structures). The essential choice is whether to connect geographically separated users to a central location or to connect terminals to local processors, each serving a cluster of users.

The rates for most new transmission services are not based on distance, as are services which use telephone-network facilities (public or private). This means that with new services there is less reason to reduce the distance over which data are transmitted. However, at present all rates are in some way volume-sensitive; the more data transmitted the higher the charge (although usually not on a simple linear scale). It is therefore advantageous in many cases to move data entirely off wide-area transmission facilities, and use only local connection (cables, LANs, etc.). The latter typically require only a one-time charge, as for example in buying and installing a cable, rather than recurring charges as is the case with wide-area services.

Transmission tariffs as well as techniques for reducing these costs are discussed in more detail under Network Design in Chapter 12.

LOOSELY COUPLED FUNCTIONS

9. *Does the system consist of a set of loosely related functions and/or of functions whose requirements are not well defined?*

Loosely related functions make up many applications or sets of applications. The amount and frequency of data flow between sets of functions determines the degree of relationship. More data flow indicates a close, or tightly coupled, relationship, while less flow is typically of a loosely coupled relationship.

When the system is loosely coupled or when the requirements are only partially defined, the best choice is a distributed-system structure, or possibly a decentralized structure whose processors will later be linked together to form a distributed system. If each set of functions is allocated to a separate computer, there is considerable flexibility to evolve as changes are needed (see question 2). This is particularly important in poorly defined applications, such as office-support systems. In applications of this kind, the initial implementations of functions can best be viewed as prototypes, even though the implementation is carried much further than in the typical prototype discussed elsewhere in this book. Office systems, as well as many other as yet poorly defined applications, are made up of clusters of loosely related functions. A distributed structure provides greater flexibility than a centralized structure to change each set of functions as needed.

MANAGERIAL STYLE

10. *Is the organization's managerial style in favor of distributed control?*

Chapters 5 and 8 discuss this aspect of system design. It is important to note that the term "decentralized" is usually applied to the management style referred to here as "distributed." However, decentralized management (or, more correctly, decentralized managerial responsibility) does not mean the same thing as decentralized information processing—which consists of free standing, unconnected systems. Calling this management style "distributed" accurately reflects the fact that it involves the allocation of certain types of responsibility to various levels of management, but with ties of reporting and control to and from top executives.

Although it is possible to provide some degree of distributed control in a centralized system, a greater degree can be provided in a distributed system. In an organization with a philosophy of distributed control, therefore, applications ought to be studied for their suitability to a distributed structure. Many applications will prove to be well suited to this approach, especially as main-line functions are increasingly computer-aided. Local control can also be provided in a decentralized structure, but this has the disadvantage of not providing for the interfunctional data flow which is so often needed.

In contrast, an organization which wishes to maintain close control over the computer-based information systems may find a centralized structure more desirable. All programming can be done by a central staff, and the management of the computer equipment can be tightly controlled. This is the method used in many organizations today, even in many which have a distributed-management philosophy. The advent of inexpensive computer equipment, especially micro-based personal computers, makes it increasingly difficult to maintain the tight control which full centralization represents. However, there may be specific applications which management believes are of such vital importance to the organization as a whole and/or which handle such sensitive data that centralized control is essential.

TECHNICAL RISK

11. *What is the degree of technical risk involved if a specific system structure is selected?*

One of the major advantages of a centralized system is that so many data-processing professionals are familiar with this approach. Many organizations have implemented applications using a centralized-system approach for a wide range of purposes: manufacturing, health care, distribution, banking, insurance, many levels of government, and others. Even if an organization has no one on the staff with personal experience in a specific application, it is often possible to hire someone who has that experience. The greater the degree of experience in the use of a particular technique, both in the local IRM staff and in the industry as a whole, the lower the technical risk involved in additional implementations using that same technique.

A decentralized system, when implemented by a professional data-processing staff, similarly represents a low technical risk, since this is, for practical purposes, exactly the same as multiple centralized systems. (In fact, the risk may be lower because of the lower degree of complexity in a decentralized approach.) If, however, the decentralized applications are implemented by programmers who are users rather than DP professionals (as is sometimes the case in decentralized systems), the technical risk is higher. Typically, however, the risk in this situation is lower than if a similarly inexperienced staff attempted to implement a centralized system involving multiple applications.

For most organizations today, a distributed system represents a higher technical risk than either a centralized or a decentralized system. There are many IRM departments in which no one has had any personal experience with systems of this type, and there is not yet a large enough pool of experienced personnel in the industry to make hiring a simple solution to this problem. Over time, the degree of risk involved will diminish steadily as the level of collective experience rises.

If there is no local experience with distributed systems, it may be possible to minimize the risk by using prototype implementations as a learning tool. It may also be appropriate to define a more loosely coupled system structure (see question 9) than might otherwise be chosen, because loosely coupled distributed systems are very much like decentralized systems. The more tightly integrated the parts of a distributed system, the greater the complexity of the design and implementation and the more effort required to keep this complexity under control. If the IRM staff is inexperienced in managing complex implementations of this type, it is a good idea to minimize the problems until the staff has acquired the necessary experience. In fact, it is wise to choose loosely coupled structures rather than tightly integrated ones whenever feasible, to minimize complexity and maximize flexibility.

COST

12.   *Will a distributed-system structure result in lower total life-cycle costs for the system?*

Placing cost last in this list of questions does not mean that it is the least important; cost is always of great importance in system design and implementation. Even if the answers to earlier questions definitely point to one type of structure, cost must also be evaluated. This can be done only by formulating some hypothetical system structures and estimating the cost of each. A detailed cost analysis can seldom be made at this point in the design process, even for a single approach. But estimates can be made, and if they are made consistently for all possible approaches, this is a workable way to make a decision based on cost.

All relevant costs must be included in this analysis and must be evaluated in terms of the expected life cycle of the system. The costs to be considered were described in Chapter 2, in the context of defining an application's potential return on investment. By this time in the analysis and design process, it is generally possible to prepare a more accurate estimate of costs than was feasible earlier. A briefly restated list of costs to be considered follows:

- Hardware costs

- System-software costs

- Application-software costs

- Communications-facilities costs

- Maintenance costs (hardware, software, etc.)

- Supply costs

- Training costs

- Facilities costs (computer room, building services, etc.)

- Operational costs

- End-user costs

Although it is difficult to generalize, total costs are lower in some cases if a distributed structure is chosen. This is particularly true if the cost (or potential cost) of communications facilities in a centralized system can be minimized by implementing some functions in local satellite processors or terminal controllers. The total cost of those processors or controllers, including program development and maintenance as well as hardware-related costs, must of course be realistically evaluated. There are techniques which can be used to reduce communications costs in a centralized system; for example, the use of public-data-network or value-added-network facilities instead of private links, or the use of multiplexors. These costs will, however, typically still represent a significant portion of total system costs in a centralized approach.

In some systems, the comparison of expected costs among different possible system structures has resulted in a somewhat higher cost for a distributed system than for a centralized system. However, some organizations have chosen the distributed alternative in spite of its apparently higher cost, to achieve other advantages which were indicated by answering the other questions listed in this chapter. One evaluation of this kind produced an estimate of 10 percent higher life-cycle costs for a distributed than for a centralized system. In that application, the reduced vulnerability to failure provided by a distributed system was determined by management to be worth the extra cost. In another case, the improved flexibility for relatively easy expansion in a high-growth situation was the determining factor in selecting a distributed system, even at added cost. Decisions of this type are unique to each organization and set of applications.

SUMMARY

The twelve questions in this chapter can help in deciding which information-system structure to choose. By answering these questions, a system designer ought to be able to decide whether a centralized, decentralized, or distributed system will be better suited to the application(s) being analyzed. This is a strategic-level design decision and must be followed up by more detailed decisions, such as where to allocate functions if a distributed or decentralized approach is chosen. (Those decisions are addressed in *The Distributed System Environment*, referenced earlier.)

The preceding questions and the discussion which accompanies each also provide information on the potential advantages and disadvantages of each system structure. Because these advantages and disadvantages are so important in system design, they are summarized here for easy reference.

CENTRALIZED SYSTEMS

Centralized systems have the following potential advantages and disadvantages (the term "potential" is used because any specific advantage or disadvantage may not be germane to all applications or all organizations).

Advantages

- Supports tight centralized managerial control
- Simplifies the protection of security or privacy
- Low technical risk because of wide experience with this approach

Disadvantages

- Lack of flexibility for change
- Difficult to expand capacity rapidly

- Lack of responsiveness if large number of terminal users must be served
- High availability resulting in high cost
- Vulnerable to single-point failures (such as in the environment of the computer center)
- Difficult to provide customization
- High cost of data communications if users are geographically remote or located at a large number of sites

DECENTRALIZED SYSTEMS

Decentralized systems have the following potential advantages and disadvantages:

Advantages

- Flexibility for change
- Responsiveness to large number of terminal users
- High availability achievable at optimum cost
- Relative invulnerability to single-point failures (although specific functions may be vulnerable)
- Support customization
- Lower cost of data communications than with a centralized structure if users are geographically dispersed
- Ability to mirror fully decentralized management control (which is a relatively unusual condition)
- Low technical risk

Disadvantages

- Difficult to support rapid expansion of capacity (same problem as in centralized systems)
- Vulnerable to intrusions into secure or private data or functions (because computers are typically not in protected rooms)
- Lack of relevance to data flow in most organizations (few applications are completely freestanding and independent from other applications)

DISTRIBUTED SYSTEMS

Distributed systems, the third choice, have the following potential advantages and disadvantages:

Advantages

- Flexibility for change

- Supports rapid expansion of capacity

- Responsiveness to large number of terminal users

- High availability achievable at optimum cost

- Supports customization

- Lower cost of data communications than with a centralized structure if users are geographically dispersed

- Ability to mirror distributed management control (the most common organizational philosophy today)

Disadvantages

- Potential for "anarchy," especially if tightly centralized control is desirable

- Vulnerable to intrusions breeching security or privacy

- Technical risk caused by lack of collective, industrywide experience

To summarize the potential advantages and disadvantages of the three types of system structures: For complex information systems the distributed-system approach offers more potential advantages and fewer potential disadvantages than either of the other structures. One potentially serious drawback of this approach is the lack of widespread experience with distributed-system implementations. Lack of experience leads to technical risk and may also result in higher costs than estimated, because of the inability to accurately forecast the effort required for design and implementation. In addition, there is some possibility that the implementation will fail, either because the design is too complex for the skill of the IRM staff or because the system design was poorly chosen through inexperience.

The risk involved does not mean that distributed systems must always be avoided. The level of risk decreases with increased experience, and the only way to gain experience is to implement systems of this type. As this volume points out repeatedly, the majority of all future computer-based information systems will use a distributed structure, because of the need to mirror complex, interconnected organizational relationships. Recognizing the degree of risk allows the successful management of that risk. Some methods of risk management are described in this chapter, including (1) using prototype implementations to gain experience, (2) implementing a decentralized structure and later linking the components to form a distributed system, and (3) minimizing the complexity of the distributed-system structures and relationships used until the staff has gained experience in working with simpler distributed systems. Although risk is involved, the choice of a distributed system will usually be the correct choice for a complex information system.

11

DATABASE DESIGN

Database design is a key element in a complex information system, because the database in many ways represents the focal point of any computer-based system today. This chapter discusses how to decide what the structure of the system (or global) database will be; that is, how the entire set of data needed by the application(s) will be organized and where it will be located within the system.

The design of the database is closely related to the other aspects of system design which are covered in Chapters 9, 10, 12, and 13. Chapter 9 discusses the design of user interfaces, emphasizing interactive methods. User interfaces affect database design because the users' need to access specific types of data, with stated response speeds and frequencies, determines many of the data elements to be stored and also influences which organizational method(s) are chosen and where parts of the database are located.

Chapter 10 describes how to choose an information-processing structure—centralized, decentralized, or distributed. That choice is very closely related to database design and determines the range of possible database structures. If, for example, a centralized approach is chosen, the database must also be centralized—although it can be partitioned to form multiple areas or independent databases, or perhaps one or more databases plus multiple files. Strategic-level system design is of course an iterative process, and an earlier decision to centralize the system may be reevaluated if the analysis of database design indicates that distribution would be advantageous.

Chapter 12 discusses the design of the network which links the computing and terminal elements of the system. Wide-area networks, made up of public or leased circuits, represent a significant recurring cost in many systems. The location of databases or database partitions can have an important effect on network costs, so the relationship between the design of the database and of the network must be considered. Chapter 13, the final chapter in this section on system design, discusses how to design for integrity (including security and privacy) and for flexibility. Both of these attributes are of great importance to a database, so preliminary design decisions made in the process described in this chapter may be modified later to improve flexibility or integrity.

This chapter provides some guidelines for good database design. It also suggests methods for partitioning the system database into manageable segments and ways to decide where to place the resulting segments.

A few definitions are important before beginning the discussion of database design. A *database* is a generalized collection of data belonging to an organization, company, or installation rather than to an individual. A database is typically organized to reflect data relationships which exist in the real world; thus the database structure mirrors (some subset of) reality. A database is usually shared by many applications, and its length of existence is independent of the execution of any particular application or application program.

A *distributed database* is a single, logically related collection of data which is either segmented or copied and attached to more than one information processor within a distributed system. If the logical database is separated into nonredundant sections and spread across two or more information processors, it is called a *partitioned database*. An example partitioned database is shown in Figure 11-1. In this context the term "nonredundant" means that any data element exists in only one partition of the database. If, instead, all or parts of the database are copied at two or more locations, this is a *replicated database*. An example of this type of structure is shown in Figure 11-2. In practice, most distributed databases include both partitioning and replication. (Distributed database structures and their use are covered in more detail in the earlier volume, *The Distributed System Environment.*)

The total set of data stored by the system may include files as well as one or more databases. A *file* is a collection of related data which has a relatively simple structure (e.g., sequential, indexed sequential). Files may belong to the system in the same way that databases do or may be private and belong to a single user, user group, or application. Many office-support functions use files rather than databases for the storage of documents, personal or group schedules, and similar information.

DESIGN GUIDELINES

Although the database design for each information system is unique in at least some aspects, the following guidelines apply in all cases. These

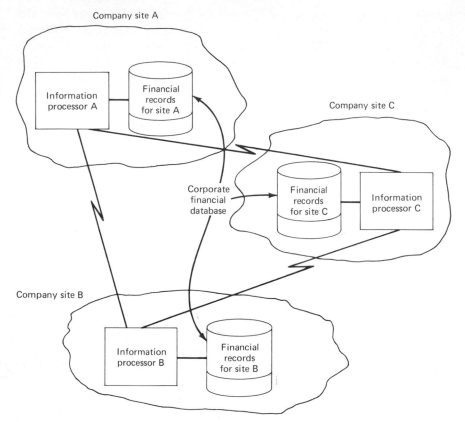

Figure 11-1   Partitioned database.

guidelines have been formulated from the experience gained by a large number of people in implementing a great many databases. Designers of databases will find it useful to keep these two rules in mind throughout the analysis and design process.

1. *Applying* **grand design** *concepts to database design, with the result that all possible data relationships are expressed in a set of exceedingly complex linkages, is usually a serious mistake.*

   This statement may be viewed as heresy by the proponents of network-structured databases, whose philosophy is that the database structure models the real-world relationships of the organization. The structures and data relationships of most organizations are very complex and if modeled completely, result in databases with two potentially serious drawbacks. First, it is often difficult to provide rapid access to data, because of the complicated data structures. Second, it is extremely difficult—sometimes, for practical purposes, impossible—to change such complex structures. Even through the database management system (DBMS) software may be able to keep track of dozens of relationship linkages among data elements, the database designers often do not really understand all the subtle relationships and therefore are typically unable to anticipate all the effects of change. In this situation, any modifications to the structure can quickly lead to chaos.

These comments must not be taken as a condemnation of network structures (with which the author first worked in 1964) but rather as a caution not to be carried away by their capabilities—or by the capabilities of any type of DBMS. There is great elegance in simplicity, and in database design—as, indeed, in all aspects of information-system design—the simplest structure which can accomplish the stated objectives is the best.

Rather than designing a *corporate database* which embodies all the data elements and relationships of the organization, it is better to design a series of loosely coupled databases. Each database can be designed to serve a functional area (marketing, engineering, etc.) and the related applications, or possibly each can be organized to serve a major set of

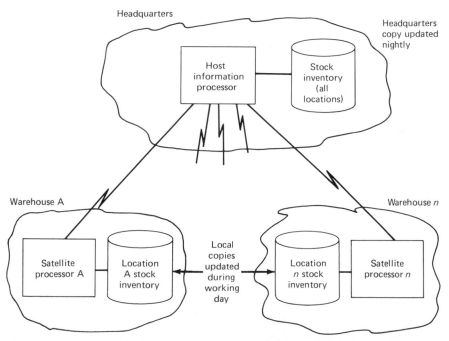

Figure 11-12   Replicated database.

applications. Data transfer among those databases will probably be required, and some redundancy of data elements may result from this approach. These consequences are preferable to the potential disadvantages of a fully integrated database. The offsetting advantages are greater flexibility to change the structure as necessary and more ability to tune the data structure for quick access.

Although this chapter concentrates on database design, many information systems include data which can be very adequately handled in file structures. If a file structure, such as sequential or indexed sequential, can provide the necessary facilities, it is a better choice than a database structure. However, a simple structure must not be chosen without a complete analysis of user requirements. If ad hoc access to the data is a major requirement or is likely to become one at a later date, it may be necessary to select a more complex structure to meet this need.

2.   *Plan for flexibility, with more emphasis in database design than in any other aspect of the system, because changing the structure of a database is usually one of the most difficult tasks in a computer-based system.*

The need for flexibility and the difficulty of achieving it in complex database structures has already been discussed. It is also covered in more detail in Chapter 13 and is stated here only for emphasis.

PARTITIONING THE SYSTEM DATABASE

If all the data elements to be stored are considered a system (or global) database, it will rarely be practical to manage them all in a single integrated database. The volume of stored data used in computer-based systems is increasing rapidly; in fact, the rate of growth in storage devices attached to large computers in the United States is currently 40 to 60 percent a year, compounded. Over time, therefore, it is becoming increasingly likely that some form of segmentation will be required. This section discusses how to partition the total set of data elements to be stored; the following section describes how to decide where to place the resulting segments within the geography of the entire system. Partitioning decisions can be simplified by answering the following questions.

1.   *Are there data-access patterns reflecting the different data requirements of user groups (including management users, if applicable) which can indicate how to partition the database?*

As the discussion in Chapter 8 points out, when data-access patterns exist, they supply a convenient way to partition the system database. In Chapters 7 and 8 the data-access patterns for management and for other users (such as those entering transactions routinely during the course of their duties) were discussed separately. All the resulting patterns relate to data access and can be considered as possible ways to partition the database.

The sample partitioned database in Figure 11-1, shown earlier in this chapter, resulted from partitioning on data-access patterns. Each of the sites shown is a major subdivision of the corporation and each maintains its own set of financial data which is unique to that location and accessed by the local financial staff and management. Until quite recently, these partitions formed three independent databases. When consolidation of corporate financial data was required, printed reports were prepared and mailed to headquarters for reinput to consolidation programs. Linking the three sites by a communications network has transformed the three databases into a partitioned database (loosely coupled). Consolidations can now be effected by transmitting the necessary data elements. In addition, properly authorized corporate executives and financial staff members can immediately access all financial information at all locations.

Some types of data-access patterns do not lend themselves to database segmentation; for example, patterns which overlap heavily, so that there are no clear boundaries between data elements used by one group and those used by other groups. This relationship of patterns is found in applications such as manufacturing, in which a parts-explosion (bill of materials) database is accessed in a variety of ways by different groups of users. Overlapping makes it impractical to partition the database in terms of access patterns. Other data-access patterns may also be unsuitable for use in partitioning. For example, nonoverlapping patterns may exist, but each may result in a set of data too small to make it worthwhile to partition the database into such small segments. Unless each partition includes at least several hundred thousand bytes of data, partitioning may be possible but impractical.

If the data-access patterns discovered during pattern analysis seem to be a suitable base for segmenting the database, questions 2, 3, and 4 provide further insight into how this segmentation can best be accomplished. If data-access patterns are unsuitable for use in partitioning, the answers to questions 5, 6, and 7 may indicate other methods of segmentation.

2. *Are there patterns which separate the data elements in terms of when access occurs and which type(s) of access are needed?*

The information required to answer this question was obtained during the pattern-analysis process (see Chapters 7 and 8). If data elements can be separated into those accessed during the business hours, perhaps for production-mode transaction processing, and those accessed during the remainder of the day, perhaps for batch processing, this may provide a method of segmentation. This type of partitioning is less often usable than partitioning by data-access patterns, because usually the same data elements are accessed at different time periods but in different modes. However, if this is the case, it may lead to a different method of segmentation. Sometimes, for example, a set of data elements is accessed interactively during the business day—for this type of access, a specific data organization will be required to provide rapid response. If the same set of data elements is accessed for batch processing at night, a different organizational method may be more efficient. In cases such as this, the possibility exists that a replicated database structure will be the best choice, with the main or master version used for batch processing and a copy of the same data elements used for interactive processing (or vice versa).

A replicated database based on this type of pattern is shown in the illustration in Figure 11-2 above. A master database of all inventory items in stock is maintained on the host processor at the company's headquarters. A satellite processor is installed at each of the company's several warehouses in different cities. Each satellite processor maintains a copy of its own stock inventory, and these copies are used during the day for both inquiry and update applications. When an out-of-stock situation occurs, a query is sent to the host, which determines where stock is likely to be located and routes the transaction there. Because the host's database is not updated in synchronization with the satellite copies, balances are out of date but are accurate enough to be useful for rerouting orders.

At night all the day's transactions are posted against the host's database, which then matches the satellite copies. Batch processing against the master database includes financial reports as well as exception reports on order trends and items which are out of stock at all locations. Because the master database and the copies are used for different purposes and in different modes, the structures are different—both are network structures, but the relationships included are different. Record formats are also different, because the master carries multisite inventories while the satellites carry only a single site each.

A periodic synchronization of the database content ensures that the master and the copies do not drift apart because of subtle differences (or possibly errors) when the same transactions are applied at two locations.

3. *Are there geographical patterns based on combinations of user data-access requirements and user locations?*

As Chapter 8 points out, this is one of the best possible patterns for database partitioning. If each group of users is geographically separate from other groups and has uniquely identifiable data-access needs, those data elements form a definable database partition. This is a particularly appropriate pattern for partitioning, since the separation of each group's data from those of other groups will make change and evolution easier by decreasing the possibility of unintended effects when change is necessary.

The first choice, with a pattern of this type, is to create a distributed, partitioned database organized around the patterns, as in the example in Figure 11-1. However, other factors—such as the cost of disk files at distributed locations or the potential problems of security, privacy, or integrity protection there—may make a centralized database a better choice. In that case, partitioning is probably still appropriate, but instead of a distributed database, the patterns can be used to separate the areas of a centralized database. If the patterns reflect totally independent data needs of the different user groups, multiple independent databases will result.

In some cases the geographical, data-access, and time-and-method-of-access patterns (questions 1, 2, and 3) are all consistent, and in this case they form a good base for partitioning. As in the geographical-pattern case, the first choice is to create a distributed database partitioned in accordance with the patterns. In other situations, either a centralized database may be created with separate areas, or several independent databases may be created, all attached to the same central processing system. Another choice, as in the example in Figure 11-2, is to create a replicated database in which copies of the required data elements serve the user groups.

4. *Are there patterns of production access to specific data elements, as contrasted to patterns of ad hoc access to other data elements?*

These patterns, like those described in question 2, occur infrequently, and when present may not form a suitable base for partitioning; however, they ought to be evaluated. A pattern which separates data elements used in production mode from those used for ad hoc access may indicate the need for different data-organization methods. This is true even if the patterns overlap—or in some cases even if all the data elements exist in each pattern. In the latter two cases, this may lead to the formation of a replicated database, with the master version used for production access and the replicated copy or copies used for ad hoc queries. If ad hoc access is limited to query mode, with no updates performed, this is a relatively easy database structure to manage.

5. *Are there portions of the database which require a higher level of integrity protection than other portions?*

Segmentation by the level of integrity protection needed is seldom viable if the total system database consists entirely of data elements of the type typically associated with use by application programs. However, many new systems include the storage of large amounts of textual information as well as data. In these cases, the separation of text from data will allow the concentration of integrity protection on the data, with a lower level of protection (and therefore presumably lower cost) for the text. The partitioning of the total set of system data in terms of text and data, when possible, is a good rule. This does not mean that stored text requires no integrity protection, but rather that it usually does not require the extremely high level reserved for data stored as the only copy of the records of the business, governmental, or other organization.

These suggestions do not imply that the integration of word processing and data processing is a mistake or is unnecessary. The integration of methods, terminal devices, and certain access modes is very appropriate. It would be foolish to install two sets of terminal or work-station equipment so that one person could enter transactions into a data-processing system and also create memos, letters, or other text. The integration of the work-station equipment and—to the greatest degree practical—of the terminal-usage methods will improve efficiency as well as lower equipment costs. The ability to merge text and data, creating documents—or perhaps a database—which include both, is extremely important in many of today's applications. The use of common network facilities and common storage hardware (if not always common storage-management software) also lowers cost and simplifies the entire system structure. All these ought to be goals in new computer-based systems, as most will include the requirement to handle both data and text—if not initially, then at some future time.

Realistically, however, there are differences in the handling of text and data (although these differences do not apply in every case and must not be overstated). Text does not usually require such a high level of accuracy (integrity) as data. There are some exceptions to this rule, such as systems in which legal documents are created; in those cases the required integrity level may be the same as in many data-processing applications. The methods by which data and text are retrieved from storage typically differ; data records are most often retrieved by a specific identifier, while text may be retrieved associatively. (In some cases, of course, data records are also retrieved using associative methods.) Data records are often stored in databases with relationship linkages to other records. Text records are often stored in library-index systems, in which each document has an identifier and the identifiers form an index for document retrieval. If differences of these types exist, they form a good base for partitioning the total set of stored data.

Segmentation on the basis of integrity-protection requirements may lead to the formation of separate databases, or of a database and one or more files if the data elements are quite independent. Multiple areas within a single database may also be a good choice. In some cases this may lead to the formation of a distributed, partitioned database—although this is not always required—and other aspects, such as geographical patterns, may be the determining factor.

6. *Are there data elements which require a higher level of protection, for security or privacy reasons, than other elements?*

The separation of data by different levels of protection is a basic security- and privacy-protection method; this is the approach used by the U.S. government in assigning security classifications such as top secret, secret, confidential, and so on. If patterns of this type were discovered in the pattern-analysis process, they may form the basis for database partitioning. Data elements which require a high level of protection can be placed in a separate database partition, and the mechanisms for protection can be concentrated on that partition. If multiple levels of protection are required, it may be best to create a separate partition for each level, so that the appropriate protection methods can be applied in each case.

Protection methods used for high-security or very private data may include physical isolation; those data elements may be maintained on storage devices in guarded, locked rooms. It may be impractical to provide remote-terminal access to data with this level of security. For example, in many military systems only directly connected work stations are allowed to access top-secret databases, and those work stations are located in guarded, locked rooms just as the data-storage devices are. Although this level of protection is not often needed for business data, the segmentation of protected data from other data may be a practical technique.

7. *Are there patterns of expected change which will affect only specific segments of the database?*

Even though, in practice, patterns of change are not often useful in segmenting a database, they ought to be analyzed. Of course, separating the data elements most likely to change from the other data elements can facilitate change when this becomes necessary. If the data elements which are expected to change form a logical segment of the database, this pattern can be used for partitioning. For example, if new methods are expected to be introduced into the order-processing department, the records associated with orders may be a good candidate to form a database area or even a database partition within a distributed database. In most cases, unfortunately, expected change occurs at a more detailed level and affects only fields within records; in these cases change is not a useful mechanism for partitioning the database.

WHERE TO PLACE THE DATABASE PARTITIONS

The answers to the preceding questions will usually result in a tentative partitioning of the system database into two or more segments. In a few cases, partitioning will not be feasible because the data structures are too integrated to be separated. (As discussed earlier, one data structure which resists partitioning is a bill of materials.) Most data structures can, with care, be separated into segments.

If the database cannot be segmented, it must be created as a centralized database. If segmentation is possible, there are three choices of where to place the resulting segments:

1. Retain all partitions at a central location, attached to one information processor.

2. Separate the partitions and attach them to different information processors, either at different locations or at the same location.

3. Use the partitioning to create a replicated database, usually to be attached to multiple information processors.

Deciding where to locate the database partitions is very closely related to the decisions made in Chapter 10 about where to locate application functions. The following questions indicate how to relate the system-structure decisions to the choice of where to place database partitions.

1. *Was a centralized structure chosen in the process described in Chapter 10?*

In a centralized system, a centralized database is the only possible choice. However, the total set of data can be partitioned into areas or even into two or more independent databases if the analysis of partitioning possibilities indicates that there are few relationships among the different partitions. In some cases, parts of the total database need not use a database structure but can be organized into files attached to the central computer system.

Sometimes a decision to centralize may require reevaluation after the database requirements are analyzed. If the database is extremely large, perhaps containing hundreds of billions of bytes of data, it may be impractical to attach that much data to one information processor. There are often problems in providing rapid access to such a large database, and it may also be difficult to protect, using techniques such as dumping the data to magnetic tape. In this situation, it is probably better to partition the database and attach the partitions to two or more information processors. If the centralization of computing facilities is an organizational goal, then all the systems and database partitions can be located at the same computer site. This centralization, of course, creates vulnerability to disasters which may affect that site, as discussed in Chapter 13, and it, or other considerations, may lead to a revised decision to distribute the processing as well as the database partitions.

2. *Was a decision made during the process described in Chapter 10 to distribute the information-processing functions?*

In that case, a distributed database is usually the best choice. There are not many processing functions which can be successfully distributed without the support of database information, which in effect forces the use of a distributed database. In fact, once the decision to distribute processing is made, that is usually a de facto decision to distribute the database. The functions which are distributed also generally determine how the database will be distributed, since specific functions require access to specific data elements.

This pattern must, however, be analyzed to determine if there are parts of the database which require special attention. For example, if certain data elements require a high degree of protection for security or privacy, it may be necessary to establish the database partition which contains those elements at a central location—or at least at a location which can be physically protected. It is also essential to investigate how integrity protection will be provided at remote database partitions, especially if a partitioned-database structure is chosen. The most common method of integrity protection is to dump the contents of the database periodically onto magnetic tape and retain the tape copy (usually at an off-site location) at least until another dump is made. If database partitions are attached to processors at remote sites—such as warehouses, bank branches, hospitals, and so on—it may be both impractical and expensive to configure a magnetic-tape handler at each of those sites. In addition, remotely located processors are usually operated by personnel who are not trained computer operators and whose duties are practically limited to turning on power, calling a central location to report trouble if necessary, and possibly loading printer paper and removing reports from the printer.

One potential solution to the problem of protecting data integrity is to choose a replicated-database structure, rather than a partitioned one. In a replicated database, the master version of the data can be retained at a central location where normal data-integrity-protection methods are used. The master version of the replicated data can also serve as a backup to the remote copy or copies, although careful design is necessary to organize the database synchronization so that backup is possible.

3. Was a decision made to use a decentralized processing approach?

If the system is to consist of multiple independent computer systems, the total database must be partitioned to match. In some applications this can be easily achieved, as the data elements needed by each decentralized computer can be easily identified and are unrelated to other data elements. In other applications, the database may not match the decentralized structure so easily. If there is a mismatch between the decision to separate the processing into multiple, independent systems and the attempt to partition the database, the decision to decentralize must be reevaluated. Perhaps the functions were not separated correctly when planning the decentralized system; in this case, repartitioning the processing functions may create a match with the database partitions. Perhaps the decision to decentralize must be changed, either to choose a centralized-processing structure, or to distribute the functions.

4. Are there database partitions which can be used in inquiry-only mode?

After the database has been partitioned, each resulting segment can be studied. If there are certain segments which will be used for inquiry only, possibly by certain groups of users and/or during certain periods, those segments may be best implemented as a replicated database. The master version of the segment(s) can be retained as part of a central database, with another copy of the segment at each location where inquiry is needed. Depending on the application, each location may need the same data (identical segments), or each location may need different data (unique segments). In either case, this pattern is very easy to handle. The master version of the data can be updated and protected exactly as in a centralized database. The copy or copies can be refreshed from the master version when necessary; daily refreshing is most often chosen, although refreshing on a weekly basis may be adequate if the update frequency is low. The master version can serve as backup to the remote copies, since they are not changed and can easily be recreated if damaged. (The remote copy or copies may also serve as at least partial backup to the master version.) In many ways, this is the optimum situation for a distributed database. Note also that it is not necessary to have a distributed system to choose this database-design alternative. The copies can be retained on the same computer system as the master version, if appropriate. This, however, may eliminate one or both of the advantages of moving the data copy close to the users—improved speed of response and lower cost of data communications.

5. *Are there usage patterns which indicate that some partition(s) will be updated frequently on-line while other(s) will not?*

If this pattern exists, it may indicate that a replicated database is the best choice. Although this situation is not as simple as inquiry-only copies (as described under question 4), it still provides a straightforward way to create data copies. The data elements which will be updated on-line can be either retained at a central location or copied at remote locations, whichever fits the overall pattern of the system better. In either case, updating will take place in only one location during the period of on-line updating. For example, if on-line updating occurs during a 9-hour business day, the replicated copy or copies may be updated during that time, while the master is idle. (The sample system in Figure 11-2 uses this approach.) When on-line updating is complete, the master version can be synchronized with the copies, and it can also be updated for other purposes if necessary. While updating occurs at the master version, the copies are not updated.

There may be partitions in the same system database which do not fit into this pattern. Some partitions, for example, may be updated randomly at all times. Those are best limited to a single copy, either in a centralized database or in a distributed, partitioned form.

SELECTING THE DATABASE
ORGANIZATION METHOD

The final decision on how to organize the database or how to organize each segment of the database is not part of the strategic-level system design; however, it is typically addressed at least tentatively during this process and will therefore be discussed briefly. There are a variety of ways to organize a database: hierarchies, networks, relations, and inverted are probably the most-used methods. Most databases which are used today for production-mode processing are either hierarchies or networks. An example of a database management system widely used for hierarchical structures is IBM's Information Management System (IMS); an example of a network-oriented database management system is Honeywell's Integrated Data Store (I-D-S). Organizational methods such as relational and inverted are most often used for query and reporting purposes and in some cases (especially the inverted form) represent copies of data extracted from production databases and organized differently.

The choice of a database organization typically represents a trade-off between speed of access and ease of access. Speed of access is most easily achieved if the data organization matches the most frequently used access paths and also if the accessing programmer has a detailed understanding of how the data relationships are structured. Ease of access is achieved by avoiding the need for the person requesting access to have any knowledge of the data organization method. This means that the person's request will have to be translated into a more specific access method by some intervening software such as a query package, which results in additional overhead and usually in slower access. Mechanisms such as inverted data structures have been designed to make it more likely that rapid access can be provided, but those methods do not always achieve their goal.

Database-organization methods other than networks and hierarchies have not yet succeeded in providing rapid access to very large databases. (Of course, using one of these methods does not automatically guarantee fast access, and often a lengthy tuning of the database structure is required before adequate performance is achieved.) One or both of these methods will probably prove to be the best choice for large databases or partitions used primarily in transaction-processing mode.

Ad hoc access to a database of this type usually requires different software from production access, since the ad hoc user typically does not wish to navigate through the database structure looking for the required data—and in any case he or she probably does not understand the data relationships. Query software can provide the necessary mapping between the user's view—which may be of a relational structure or may simply treat data elements as entities free of any structure—and the actual database structure. If frequent ad hoc queries are needed, it may be better to create a query copy of the needed data elements, very likely using a method of organization different from that of the production database.

There is no single answer to the problem of selecting a database organization. New software techniques, probably in conjunction with hardware specialized for associative searches, may in the future make the relational method suitable for all types of use; today that is not the case.

SUMMARY

As in the case of information-processing structures, each database structure has advantages and offsetting disadvantages. The following lists provide an easy-reference guide, which can be consulted during system and database design.

CENTRALIZED DATABASE

A centralized database has the following potential advantages and disadvantages (not all of which will apply in any specific situation).

Advantages

- Supports tight centralized control over data elements
- Simplifies data-protection methods for integrity, security, and privacy
- Relatively low technical risk, since there is wide experience in creating centralized databases (however, if the organization's staff does not have database experience, the risk will be higher)

Disadvantages

- Complexity of the structure as the size of the database and/or the number of relationships increases
- Difficulty of changing large, complex structures, leading to inflexibility
- Slowness of access if a large number of terminal users must access the database simultaneously
- Vulnerability to single-point failures affecting the storage devices, the computer system, and/or the environment of the installation
- Time-consuming or complex integrity-protection procedures (e.g., dumps) as the size of the database increases
- High cost of data communications if users who need to access the database are geographically remote or located at a large number of sites

DISTRIBUTED-REPLICATED DATABASE

A distributed-replicated database has the following potential advantages and disadvantages:

Advantages

- A master version of the data can be retained centrally for tight central control.
- Overall availability can be improved, because the master version can be used to recover from errors or failures in the remote copies; it may also be possible to use the remote copies as backup to the master version.

- Fallback to a centralized database is usually possible if problems occur, lowering the technical risk of this approach.

- Access speed can be improved by locating the copies close to the terminal users; also, more processors are available to work in parallel to fulfill database-access requests.

- Cost of data communications may be reduced by moving the copies close to the terminal users.

Disadvantages

- Synchronization of the copies with the master version, especially if the copies are updated, may be complicated.

- The master version is really a centralized database and has the same disadvantages of complexity, inflexibility, and difficulty of integrity protection.

- Replication of hardware and data increases cost.

- Security and privacy protection for the copies may be difficult to achieve.

DISTRIBUTED-PARTITIONED DATABASE

A distributed-partitioned database has these potential advantages and disadvantages:

Advantages

- Data redundancy is minimized, reducing hardware cost (compared with a replicated database).

- Access speed can be improved by locating the partitions close to the terminal users.

- Cost of data communications may be reduced, because the partitions are close to the terminal users.

- No data synchronization is required.

- Loosely connected partitions may simplify the structure of each partition and also of the entire database.

Disadvantages

- Technical risk is relatively high, because few organizations have any experience with implementing partitioned databases.
- Security and privacy protection for remotely located partitions may be difficult to achieve.
- Remotely located partitions may make integrity-protection procedures complicated and/or costly.

Analyzing these contrasting advantages and disadvantages shows that the centralized database has the lowest technical risk initially, because of available experience; but as the size of the database grows, the disadvantages become very significant. The partitioned database has the most long-term promise, especially if the partitions are relatively loosely related to minimize complexity. However, the lack of experience with partitioned databases makes the technical risk relatively high today. Also, some data structures do not partition well and are therefore not suited to this approach. Replicated databases, in the right situation, provide significant advantages with relatively low risk but increase total system cost because of the replicated data and hardware. Some organizations have found that choosing a replicated database is a good first step in moving toward a partitioned approach, because the initial risk is lower, and this provides an opportunity to gain experience in working with distributed databases.

12

NETWORK DESIGN

When strategic-level decisions have been made on how to structure the information processing and database(s) of the new system, high-level network design can begin. The focus of this chapter is on how to select the type or types of network facilities to be used; how to decide which network structure is appropriate; and, generally, how to minimize the cost of the network. An initial network design often results in unacceptably high cost, and other alternatives must then be considered. Because of the enormous variety of different network facilities available in different countries, and even in different locations within the same country, this chapter does not attempt to provide an in-depth procedure for network design—which, in any case, is not part of the strategic-level system design covered in this book. The methods described in this chapter can be used to determine the best network approach for the system being studied; more detailed design methods can then be used to select the exact types and capacities of the links and to determine the exact routes for those links. The Bibliography lists several sources of more detailed procedures for network design.

This chapter is divided into three parts, the first of which describes the types of network facilities which are available and from which the network designer must choose. Transmission facilities are described, as well as the protocols which are part of network software and also the need for compatibility among the protocols used. The second section of this chapter describes how to choose the best type or types of facilities to use, based on the system's requirements and the system-design decisions made earlier, using the methods described in Chapters 10 and 11. The second section also describes how to reevaluate those earlier decisions if an efficient network cannot be configured at an acceptable cost.

The final section in this chapter covers some facets of networking which now apply to only a few information systems but which ought to be studied carefully for future applicability. Information-processing functions are now offered in some public data networks and value-added networks. This opens up new possibilities when designing a system which requires those functions; one example is electronic mail. The emerging technique called *viewdata*, which allows a television set to be used as a terminal device to access computer-based facilities, is also described. While not many information systems make use of viewdata yet, it has very interesting potential for the future.

NETWORK FACILITIES

The problem of network design is to provide the connections necessary to allow data transfer between any of the system components (information processors, terminals) which need to communicate. Network facilities which provide the needed transmission can be categorized into five groups: four of these are wide-area network facilities; the fifth covers local-area networks. A network to support a computer-based information system may consist of any one type of facility or any combination of the five types. The five classes, each of which is described below, are the following:

- Private links

- Private meshed networks

- Public switched networks

- Public data networks and value-added networks

- Local-area networks

PRIVATE LINKS

Private links are transmission facilities obtained from a common carrier* in the United States or from the postal telephone and telegraph (PTT) authority in most other countries for dedicated use by the leasing organization. The advantage of a private link is that the connection is always available and no time or action is needed to establish it when data are to be transferred. This contrasts with the use of a switched network, in which a connection must be established each time data transfer is required and contention for use of the network facilities may make it impossible to establish a connection when it is needed.

---

* An organization authorized and regulated by the Federal Communications Commission (FCC) to provide transmission services to the public.

Private links can be configured in two ways: point-to-point and multipoint. A *point-to-point link* connects two endpoints, usually a computer and a terminal device or terminal controller, but it may also link two computers or two terminals. A *multipoint link* connects three or more endpoints and may be more economical than a point-to-point connection. Multipoint links are commonly used to connect several terminals to a computer, but there are also situations in which a multipoint link is used to connect several computers together. Both point-to-point and multipoint link configurations are shown in Figure 12-1.

Point-to-point private links can be obtained in different speeds, which is the common way of referring to the transmission capacity of the link. Available speeds range from subvoice grade (less than 2400 bits per second, and seldom used in the United States today) to over 1 million bits per second. The most commonly used speed in the United States is voice grade, which has a nominal capacity of 2400 bits per second. However, with suitable modems a voice-grade link can transmit data at speeds up to 19,200 (19.2 K) bits per second, assuming that the devices at each end of the link can operate at that speed. Transmission speeds have been increasing over time, both because of more use of computer-to-computer transmission and because many terminal devices today can effectively use high-speed links. When only teleprinters were used as terminals, even a voice-grade link was able to transmit data faster than the printer could handle it, so a higher speed was unnecessary. Video-display terminals, especially those which provide graphics capabilities, can effectively use speeds as high as 19.2K bits per second (K equals 1024 bits). Some terminals use asynchronous transmission methods, in which

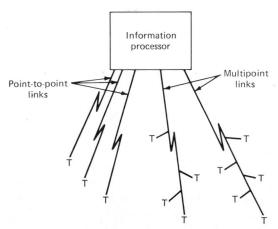

Figure 12-1 Point-to-point and multipoint links.

each data character is transmitted independently, surrounded by synchronization bits. Most contemporary VDTs transmit data synchronously; in this method data characters are sent in a continuous bit stream, and the sending and receiving devices synchronize only at the beginning of each transmission. Synchronous transmission can achieve higher effective speeds than asynchronous and can therefore make better use of high-capacity links.

Links with speeds greater than 19.2K bits per second are called *broadband* and may allow for transmission at 48K, 56K, 72K, or even more bits per second. These are the most commonly used speeds, largely because the cost of a link (in the United States) rises very rapidly above 72K bits per second. Some systems use 1.5-megabit-per-second links to connect large computer centers, and there may be installations using even higher speeds (which are technically feasible but prohibitively expensive for most applications).

Channels on communications satellites also allow for very-high-speed transmission, although the speed at which each user of the satellite sends data also depends on the services and rates offered by the organization which controls the satellite. Some services allow for the leasing of a satellite channel to operate at voice-grade speeds as well as at much higher speeds. (One of the major uses of satellite channels today is for the transmission of television programs, which require very wide bandwidth—effectively the same as saying that they require very high capacity or speed.) Satellite transmission is inherently multipoint, in that any transmission beamed to a specific satellite (of which there are a number in orbit) can be received by all the earth stations tuned to that satellite. Some applications take advantage of this fact to broadcast messages or data to multiple stations via the satellite. In most data-processing applications, however, any transmission is logically aimed at a specific endpoint via the associated earth station.

Other land-based (terrestrial) communications links can be specifically configured to provide a multipoint connection. This configuration is most often used to link a set of terminal devices to a central computer or computer complex. The rationale behind the use of multipoint links is that each terminal device (or cluster) requires transmission to or from the computer for only a small proportion of the total available time. Dedicating an entire link to each terminal is therefore not cost-effective. Of course, sharing the link among several terminals means that the link-control protocols must resolve the contention for use of the link. Typically the computer (or its front-end processor) polls each terminal in sequence and accepts input from each one that is waiting when the device is polled. When output for a terminal is ready, the computer selects the appropriate device and transmits the data. Circuitry in the terminal devices (or controllers) ensures that only the polled or selected terminal will transmit or accept data during that time.

PRIVATE MESHED NETWORKS

Private meshed networks are also made up of private, leased links, but the network configuration is far more complex than point-to-point or multipoint. The term "meshed" is descriptive of the typical network configuration in this category, as shown in the example in Figure 12-2.

The prototype of many meshed networks is the ARPANET, which was designed and installed by the Advanced Research Projects Agency of the U.S. Department of Defense, with the assistance of the Bolt, Beranek, and Newman Company (BB&N). ARPANET connects a large number of computers and terminals together, and because the requirements for data exchange between pairs of locations cannot be predicted, the network connects all endpoints to all other endpoints. Because traffic patterns and volume are unpredictable in ARPANET, the network provides multiple paths between any two endpoints. This allows the transmission load to be leveled across the available facilities as well as providing backup in case one or more links fail.

Because a meshed network provides multiple possible paths between endpoints, network-switching processors are required to manage the

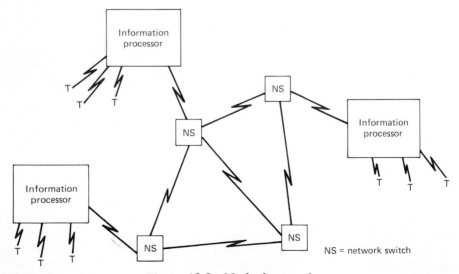

Figure 12-2 Meshed network.

flow of data within the network. Switchers of this type were first implemented using minicomputers (in ARPANET the original switchers were Honeywell Series 16 minis), but today switchers are often microprocessors. In some cases, clusters of micros are used to manage the network. The software in the network switches includes routing algorithms to determine how data are sent through the network, flow control to ensure that no part of the network is overloaded, and automatic rerouting or retransmission to handle errors and link failures.

The transmission technique generally used in meshed networks is packet switching, in which logical messages to be transmitted are broken up into fixed-length packets, usually only 80 to 100 characters in length. Segmenting messages into packets allows quite efficient use of the network links, because packets from different messages can be interspersed and short messages such as transactions need not wait for the completion of long messages such as jobs or files. In some meshed networks each packet is routed independently, to provide still greater flexibility and efficiency. This technique is called *datagram switching*. The independent routing of packets causes considerable complexity in flow control, because the load on the links and switches is completely unpredictable from moment to moment. Establishing a *virtual circuit*, over which all packets of a message travel, is more common than the use of datagram in meshed networks today.

Most public data networks and value-added networks (described in the following subsection) are meshed networks and use packet switching. It is feasible today to install a private meshed network if the application justifies the expense. There are a number of organizations which will install this type of network on a contract basis; turnkey network organizations in the United States include BB&N, Tymnet, GTE-Telenet (a subsidiary of General Telephone and Electronics), and SESA Honeywell. In a number of other countries the SESA organization, which is based in Paris, provides turnkey network design and installation. There are, however, some countries in which the establishment of private networks is discouraged or prohibited by the PTT, in which case this choice does not exist.

Another category of complex network sometimes referred to in computer-industry literature is a *ring network*. As the name suggests, this is a network which has a circular structure, connecting processor(s) and terminals in a loop. In practice, ring networks do not provide a viable general solution to wide-area interconnection problems. However, they are one form of local-area network and are discussed under that heading later in this section.

PUBLIC SWITCHED NETWORKS

Public switched networks are the telephone networks used primarily for voice communication. However, a great deal of data traffic also travels on those networks. In some countries the PTT may discourage or prohibit the use of the public voice network for data or may control the possible methods of use, and any restrictions of that type must be observed.

When the switched network is used, a connection must be established each time data transmission is required between two endpoints, whether a terminal and a computer, two computers, or two terminals. It is not usually possible to connect more than two endpoints using the switched networks. The connection can be established by manually dialing (or touching) the number of the desired endpoint at the calling endpoint; this is the method often used when a remote terminal location calls a computer system. Automatic dialing (touching) may also be possible if the calling terminal or computer is equipped with the necessary hardware and software and if the organization which operates the switched network allows automatic call establishment. Considerable freedom and flexibility are typical in the U.S. networks, while restrictions of various types are found in some other countries. A careful review of local regulations and authorized methods is required in each case.

Using a switched network is often the best choice if two locations need to exchange data only infrequently and if they can afford to wait when no connection is available. Sometimes an attempt to obtain a connection results in a busy signal, either because the called endpoint is already connected to another location or because the network circuits are overloaded. Another disadvantage of a switched network, in contrast to private links, is that data-transmission quality may vary enormously. Sometimes data transmission may proceed with few or no data errors; at other times a large number of transmission errors may occur. One reason for the variability of errors on the public networks is their method of routing; a call between the same two endpoints may travel over different routes at different times, and some routes may include equipment which is more error-prone than on other routes. Also, some types of transmission facilities are sensitive to disturbances such as weather (thunderstorms, etc.), which can cause data errors during transmission. Microprocessor-based modems which are available in many markets today can help to minimize the impact of data errors by dynamically equalizing the transmission links during data transfer. These modems are typically more expensive than those which do not provide this feature, and some PTTs may not allow their use. Individual decisions must be based on the local conditions and cost trade-off studies.

PUBLIC DATA NETWORKS AND
VALUE-ADDED NETWORKS

Public data networks (PDNs) and value-added networks (VANs) are public facilities, shared just as the voice networks are shared but provided for data transmission only. The term "value-added network," or VAN, is used in the United States because the organizations which operate networks of this type obtain their basic transmission facilities from other organizations—typically the American Telephone & Telegraph Company (AT&T) or one of the independent or Bell System telephone companies—"add a value," then resell the facilities to their customers. Originally the value added was the ability to share facilities (which was not originally allowed within the United States), thereby lowering cost. Automatic error detection and correction for data transmitted over the network facilities is another value-added service. Some value-added carriers now offer new added values, such as the delivery of electronic mail; those services are discussed in the last section of this chapter.

In countries other than the United States, shared networks which offer only data services are called *public data networks*, or PDNs. The only difference between VANs and PDNs is which type of organization provides the network services; in most countries the government-controlled PTT manages the PDN(s). An exception to this pattern is found in Canada, where private companies (similar to the United States-based common carriers) provide PDN facilities. Generally speaking, all PDNs and VANs are similar, regardless of the country and the organization which operates the network. At a more detailed level, each PDN and VAN is unique, since no two provide exactly the same facilities or charge the same rates. The comments in this section are therefore general; before any serious consideration is given to the use of a PDN or VAN, it is necessary to study the available service(s) in the affected locations. From this point on, the term PDN will be used to refer to both PDNs and VANs, unless a specific distinction is necessary.

There are two classes of PDNs. One class uses packet switching and provides interconnection using the X.25 recommendation of the International Telephone and Telegraph Consultative Committee (CCITT); the other class uses circuit switching, with interconnection via the X.21 CCITT recommendation. Packet-switched PDNs are therefore often called X.25 networks, while circuit-switched PDNs are called X.21 networks. At present there are more X.25 than X.21 networks.

X.25 PDNs use the meshed configuration and packet-switching methods described earlier for private meshed networks. The PDN provides complete delivery service; that is, the subscribing organization attaches its computer(s) and terminals to the network and when data transfer is needed, the data are supplied to the network using the prescribed protocols. The network ensures that the data are delivered to the correct, addressed endpoint and that any transmission errors or problems in the network are corrected or bypassed. The correct delivery of the data, end to end, is the responsibility of the network supplier.

The X.25 interconnection definition by CCITT is used by most packet-switched PDNs; however, because the X.25 specification has a number of options, each PDN uses a somewhat different version of the protocols and provides a somewhat different set of services. The X.25 definition allows for three levels of protocols, as shown in Figure 12-3. The subscriber's hardware and software connection to the PDN must match these three levels.

There are CCITT provisions for devices to attach to an X.25 network without using the full set of protocols. Unintelligent terminals can be connected using a *packet assembly and disassembly* (PAD), which is typically microprocessor-based. CCITT recommendation X.28 defines the operation of a PAD, while X.29 defines the interface of X.25 termination devices with a PAD. PADs are therefore sometimes referred to as X.28/X.29 devices. In the United States the VANs also provide other, nonstandard interconnection methods for computers and terminals which do not support X.25, X.28, or X.29; it is expected, however, that over time more subscribers will convert to the standard interconnection methods.

Circuit-switched PDNs are very much like the public switched net-

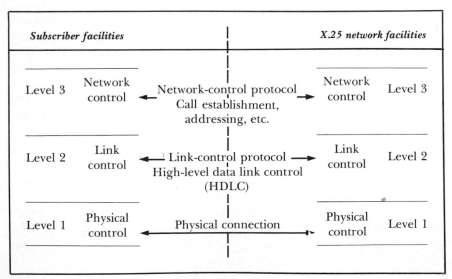

Figure 12-3. X.25 Connection

works used for voice; however, these switched networks are used only for data. Like the packet-switched PDNs, the circuit-switched networks are controlled by micro- or mini-based network-switching processors, which establish the connections between endpoints that need to exchange data. The subscriber device that wishes to establish a circuit does so by presenting a called location identifier to the PDN, which then establishes the connection and allows data transfer to occur. Unlike the packet-switched networks, the circuit-switched networks provide a dedicated link for the duration of the connection, without sharing among the different users of the network (again, this is very similar to the way in which voice networks operate). When data transfer is complete, one of the endpoints involved must break the connection by notifying the PDN to do so. The X.21 protocol used to connect to circuit-switched networks is a subset of the X.25 three-level connection but also includes call-establishment and disconnection methods which may be unique to each network.

Packet-switched and circuit-switched PDNs have somewhat different capabilities and somewhat different advantages. Packet switching is a very efficient, and often a low-cost, way to send small amounts of data (e.g., transactions and responses) among large numbers of endpoints, especially if the pattern of connections among the endpoints changes often and is difficult to predict. Packet switching is not, in general, a very good way to send large volumes of data between endpoints, especially if the endpoints also exchange interactive data such as transactions.

Circuit switching, in contrast, can be quite costly when used to send transactions and responses unless these occur continuously and, in effect, need a dedicated circuit. (In that case a private link is usually a better choice.) Circuit establishment requires some time, and the PDN supplier usually charges a fee for each time a circuit is established and broken. The supplier also usually charges for the time during which the circuit remains in operation, so that keeping a circuit in place when no data are being transmitted can be expensive. There are software techniques which can minimize the overhead and cost of using circuit switching for short messages, but in general the cost is higher than with a public switched (voice) network. On the other hand, circuit switching is very well suited to the transfer of files, jobs, or other relatively high-volume data streams, as the cost and overhead of circuit establishment are worthwhile for each transmission.

In many cases, the use of the PDN (or VAN) may be impractical because the locations where the computers and terminals will be placed are not yet served by a network of this type. Also, although the differences between X.25 packet-switched PDNs and X.21 circuit-switched PDNs may make one of these forms more desirable, many locations are served by only one type of PDN. A list of the operational PDNs and VANs and those planned for startup within the next few years is pro-

Country	Network	Type	Status
Canada	Datapac	X.25	O
Europe	Euronet	X.25	O
Federal Republic of Germany	DATEX-P	X.25	O
France	Transpac	X.25	O
Japan	DDX	X.25	O
Scandanavian countries	NPDN	X.21	O
United Kingdom	PSS	X.25	O
United States	GTE Telenet	X.25	O
	Tymnet	X.25	O
	ACS	X.25	P
Netherlands	DN1	X.25	O
Spain	CTNE	X.25	O
Australia	Austpac	X.25	P
Italy	?	X.25	P
Austria	DATEX-P	X.25	P
Brazil	?	X.25	P
Argentina	?	X.25	

O = Fully operational as of August 1982
P = Planned

Figure 12-4. Public Data Networks and Value-Added Networks

vided in Figure 12-4. Since new networks of these types are being planned continually, the list of planned facilities will become obsolete, and the network designer must remain in touch with developments in his or her own area.

LOCAL-AREA NETWORKS

Local-area networks (LANs) form the fifth class of network facilities and differ from the other four types, the wide-area networks, in being restricted to relatively short distances. (The other types of facilities can encompass distances as great as worldwide interconnection, if necessary.)

Although the term "local-area network" has been used for a relatively short time and usually refers to specific forms of local connection, systems have been connected by cables for years; this is one form of LAN.

It is convenient to separate LANs into the following categories:

1. Short-distance communications which are identical to wide-area networks except that cables instead of long-distance circuits are used (not considered to be LANs by many people)

2. Medium-distance cable or bus connections, usually of medium speed, which interconnect work stations, terminals, processors, peripherals, etc.

3. Very-high-speed bus or cable connections which interconnect processors and possibly peripherals

The first category listed above is shown in an example in Figure 12-5. In the figure, terminals relatively close to an information processor are connected to that system via cable. These connections are often multipoint but can also be point to point. Depending on the specific processor, these connections may be treated as I/O channels rather than as a communications network. Other systems treat this type of connection as a network, although one of short distance. If the distance between a terminal and the processor exceeds about 45 meters, it is usually necessary to install modem-bypass devices at each end of the cable. The modem bypasses amplify the data transmitted over the cable, so that

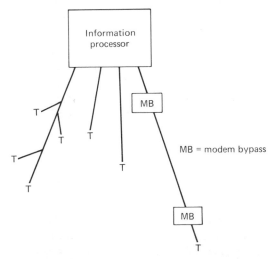

Figure 12-5 Terminal-connection LAN.

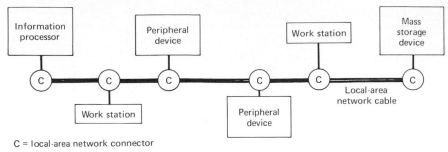

C = local-area network connector

Figure 12-6 Medium-speed LAN.

longer distances can be handled without an unacceptable diminution of the signals.

Connections of these types are very economical and also very simple to install and operate. Typically the same software which handles remotely connected terminals or work stations can handle this type of cable connection. (This is the reason that some people do not categorize cable nets of this type as LANs.) As the cost of fiber optics decreases, LAN connections of this type will probably use that technology instead of wire cables, as is typical today. Fiber optics provide extremely error-free transmission and are also very resistant to interference from radiation, corrosive atmosphere, and so on—problems which often exist in factories, for example.

The second category of LAN is more typically classified under that heading; this is the medium-distance cable or bus connection, usually oriented toward connecting work stations, processors, and possibly other devices in an office or factory environment. Ethernet, defined and implemented by Xerox, is probably the best-known example of this class of LAN.

The medium-speed LAN is typically a coaxial cable with connectors for each device to be attached to the cable, as shown in the example in Figure 12-6. Messages can be sent between any pair of devices attached to the cable. The protocols used on this type of LAN vary, depending on the specific implementation, but are of two types; those which use contention methods and those which use tokens. Contention protocols allow any device to attempt to transmit data at any time that the cable is free; if another device starts to transmit at the same time, contention results, and both devices discontinue transmission and wait for a randomly selected period of time before trying again. The delay will normally make the cable free when the first one tries again to transmit. Token protocols are somewhat more complex and involve passing a token indicating which device on the cable is allowed to transmit at a specific time.

Transmission speeds on this type of LAN range from a few thousand to several hundred thousand bits per second; higher speeds are not usually required, because of the nature of the connected devices and the types of data which they exchange. A single LAN cable is limited to about 1.5 kilometers because of signal attenuation if transmission over greater distances is attempted. This distance is long enough for many applications in the office and factory, where devices which need to exchange data are in relative proximity to one another.

Low- to medium-speed LANs are usually configured in a straight line running past all the processors and devices to be interconnected. However, some LANs can also be configured in a ring structure, so that the cable forms a loop onto which devices and processors are connected. Only certain physical arrangements are well suited to the use of a ring-structure LAN, and the locations to be connected must be relatively close together so that distance limits are not exceeded. Some specialized high-speed LANs (discussed next) have also been implemented using a ring configuration.

The third type of LAN is very similar to the second but differs in being oriented mainly to processor-to-processor connection with very-high-volume transmission. As a result, high-speed LANs may use a bus or computer I/O channel instead of a coaxial cable. Fiber optics can also be used in a high-speed LAN. In a computer installation which includes three large mainframes, a high-speed LAN might allow the computers to exchange jobs and files, and/or to access a set of shared disk devices. A schematic of a system of this type is shown in Figure 12-7.

The best-known supplier of this type of LAN is Network Systems Corporation (NSC), whose HYPERchannel is used in a number of large computer centers. The basic HYPERchannel supports data transmission

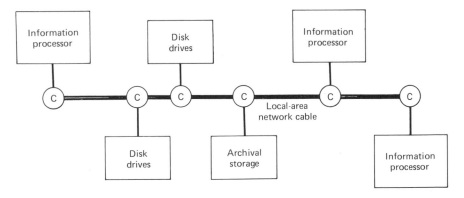

C = local-area network connector

Figure 12-7 High-speed LAN.

at megabyte speeds, and hardware adapters are available to connect many different types of computer systems to the cables; examples are IBM, CDC, Honeywell, Cray, DEC, and Univac. NSC also offers a lower-speed LAN, the HYPERbus. A five-layer set of protocols, NETEX, has been defined by NSC to be compatible with the Open Systems Interconnection standard discussed in the next section on protocols. NETEX is designed for use on both the HYPERchannel and HYPERbus LANs.

The design of LANs is still quite new, so less implementation experience with them exists in the industry than with wide-area networks. Many designers of future systems will, however, find it essential to take advantage of LANs, since these represent a very efficient and cost-effective way to connect work stations and processors.

COMBINATION NETWORKS

Combination networks which are made up of more than one of the above five types of facilities are very common. As the use of LANs increases, combination will become more common than today. Also, as the use of PDNs or VANs for long-distance traffic increases, a PDN or VAN will often be combined with other types of facilities. Combinations are typically chosen either because two or more types of facilities provide a more cost-effective solution than a single type or because one type of facility is used as backup to another type. (The latter topic is discussed in Chapter 13.)

NETWORK PROTOCOLS

A *protocol* is a set of rules for the exchange of information and defines both how the information must be formatted for transmission and the series of commands and responses to be used in the exchange. As such, it is a very important factor in the choice of network facilities. For many years the protocols used in networks were not standardized, leading to the implementation of networks using many different protocols—which were often quite inefficient. Some organizations defined local, specialized protocols which were completely incompatible with those of any other devices or organizations. Others modified generally used protocols to meet local needs. The result was a wide variety of methods for data exchange.

Today there is a great deal of emphasis on the definition and support of standard protocols, with more commonality among different types of computers and terminals than at any time in the past. This trend is most heavily supported by the PTTs and other governmental organizations which set rules for the acquisition of computer and terminal equipment.

Equipment-acquisition rules increasingly call for the support of standard protocols, and vendors are therefore under a considerable amount of pressure to conform. Individual computer-using organizations have less impact on vendors' plans but are also increasingly demanding the ability to interconnect a variety of different types of equipment. All this pressure has led to quite rapid standardization in new equipment and facilities.

One must not, however, lose sight of the fact that few installations are "state of the art"; most have equipment which was acquired several years ago and is still usable. It is probably the case, therefore, that *binary synchronous communications* (BSC) is still the most widely used communications protocol. This situation persist in spite of the fact that IBM, which originally defined BSC, introduced a more efficient, more standard protocol—*synchronous data-link control* (SDLC)—in 1974. Many users of IBM computers and terminals are gradually converting to SDLC, but an evolution of this type does not occur overnight, even when significant benefits accrue from the change.

The International Standards Organization (ISO) has defined what is probably the most widely accepted communications protocol, *high-level data-link control* (HDLC). HDLC, like IBM's SDLC, is a bit-oriented protocol designed for two-way simultaneous transmission to take advantage of the capabilities of contemporary network links. The standardization effort which resulted in HDLC was heavily influenced by IBM's early definitions of the protocol which later became SDLC, and there is considerable similarity between the two protocols. In fact, a subset of SDLC is compatible with the standard HDLC, so that SDLC can be regarded as a superset of the standard.

The other major area of effort in ISO is the definition of standards for open systems interconnection (OSI). This effort is intended to define functional structures, interfaces, and protocols for the cooperation of systems across networks; so far the structure of OSI has been agreed on and standardized. The lower parts of the OSI structure are the same as X.25 (see Figure 12-3), so that the protocols for those layers are well defined and effectively standardized. Protocols and interfaces associated with the upper layers of the structure are both more complex than those for the lower layers and more controversial, as they represent potential changes to existing vendor software which provides the same (or a subset of the same) functions. Work is continuing within ISO and supporting national standards bodies such as ANSI (the American National Standards Institute) to define those protocols and interfaces. The OSI structure of layered functions and protocols is shown in Figure 12-8.

CCITT, which is a consortium of PTT and communications carrier organizations, has been very active in the area of connection to and

Layer	Function set
Layer 7	Application
Layer 6	Presentation
Layer 5	Session
Layer 4	Transport
Layer 3	Network
Layer 2	Link
Layer 1	Physical

Figure 12-8   Open systems interconnection structure.

between networks. Their most commonly used definitions include X.21, X.25, X.28, and X.29, as discussed earlier in this chapter. Another CCITT recommendation, X.75, defines protocols for the interconnection of two X.25 networks. If, for example, an organization operates a private meshed network and wishes to interconnect that network to one operated by another organization, X.75 defines how that can be done in a standard way. The relationship between ISO and CCITT has not always been completely amicable. At one time during the definition of X.25, it appeared that CCITT would adopt a Level 2 definition of HDLC which was at variance with the ISO-standard HDLC. That did not occur, and the practical problems of defining standards which allow interconnection are forcing the standards-setting organizations to be more cooperative than at some times in the past.

The Institute of Electrical and Electronics Engineers (IEEE) is working on the definition of standard protocols for LANs. Because of the existence of several different LAN implementations—most prominently Ethernet—there has been difficulty in agreeing on a standard set of protocols. Two different methods of transmission on LANs—baseband, as in Ethernet, and broadband—each have proponents. The intent of the IEEE work is to define LAN protocols which are compatible with the open systems interconnection model, so that only the lower layers are affected when a LAN is used instead of or in addition to a wide-area network. This is a very important goal, because this commonality will limit the differences visible to application-level software; ideally, at the application-program level it ought to be transparent which type(s) of network facilities and protocols are in use.

Although older protocols, and in many cases nonstandard protocols, are used in many installations, there is an accelerating trend in new systems toward the use of standard protocols. One very encouraging fact is that essentially every major vendor of mainframe computers and/or minicomputers has now announced the support of X.25 and X.21. IBM's announcement in the United States of X.25 and X.21 support in August 1981, followed by Hewlett Packard's similar announcement in October 1981, make this support nearly universal.

Many systems do not use standard protocols at present, and some new systems are even implemented using older or nonstandard protocols because of specific circumstances. If, for example, a terminal or minicomputer exactly suits the requirements of the application but does not support standard protocols (e.g., HDLC, X.25), it may still be appropriate to select that device. However, each organization ought to set goals for gradual evolution into the use of those standards. Otherwise, over time the flexibility to connect to PDNs and/or VANs and the flexibility to choose equipment supplied by many different vendors will be lost.

SELECTING NETWORK FACILITIES

By this time, the analysis and design process will have determined where terminals will be located and also where information-processing facilities are most likely to be placed. Both of those decisions are still subject to change, especially as the cost of the system is evaluated in more detail. If, for example, terminal- and/or network-facility costs are too high, it may be determined that some users will not be given terminals. An evaluation of network costs may lead to the conclusion that a greater distribution of processing is appropriate, so that there are more information-processing locations. However, at the beginning of network design the situation is typically as shown (in simplified form) in Figure 12-9; terminal and processing locations are known. The challenge now is to interconnect all the locations which need to exchange information and to do so with optimum efficiency and at optimum cost.

The data-collection process (see Chapters 4 and 5) supplies the essential input to network design, including:

• Which endpoints need to exchange data

• How much volume, during which periods of time, will be exchanged

• What response speed is required for each type of traffic

In addition to these facts, it is also necessary to obtain information about the network facilities which are available in the locations to be served. If only a small geographical region is involved, the possible

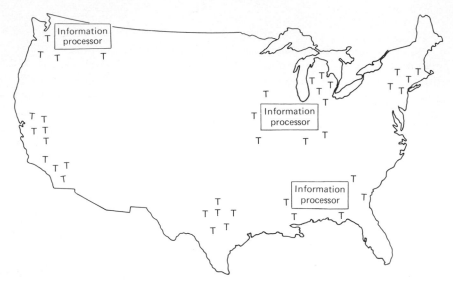

Figure 12-9   The network-design problem.

facilities may be limited; over wider areas, there may be a large number of different types of facilities. Since these represent the potential choices for use in the new system, accurate information is important.

The easiest way to obtain this information is from the local supplier of telephone services ("local" in this context refers to the organization's headquarters location). In the United States, this is either a Bell System telephone company or one of the independent telephone companies, such as those controlled by GTE, Continental Telecomm, and so on. The facilities of the local supplier will most likely be used for at least some purposes, so its cooperation may be very helpful. On the other hand, the local supplier has no incentive to make its customers aware of competing services, such as those offered by specialized or value-added carriers. If the network to be designed is a large one, spanning a wide area, investing in the services of a consulting firm specializing in network design is usually a good idea. An organization (or person) intimately familiar with the full range of available network facilities and how to combine them to achieve a cost-effective network can reduce life-cycle network costs considerably—compared with the most straightforward use of network facilities.

There are some aspects of network costs with which each system designer ought to be familiar. The first of these is the distinction in the United States between intrastate rates, which apply to links within a state, and interstate rates, which apply whenever links cross state boundaries. Rates, as defined in the tariffs approved by the FCC, for a given distance interstate are often less expensive than for the same distance under intrastate tariffs. Each state has a regulatory agency or commission which establishes intrastate rates. Because each such commission sets tariffs independently, rates vary widely from state to state. In high-tariff states, such as California and New York, it is better to use private or public-switched-network links only when necessary. If connections can be moved onto a LAN or onto interstate links, costs can usually be reduced.

The services provided by the specialized carriers, such as Microwave Communications Inc. (MCI) and by value-added carriers, such as GTE-Telenet and Tymnet, are typically lower in cost than comparable services provided by the full-service carriers. If alternative services are available in the geographical areas to be served, they are therefore worth consideration. Of course, the bottom-line costs are not always lower with alternative carriers than with the full-line carriers, so each situation must be evaluated individually.

LANs have an enormous cost advantage over any other type of facility, because they involve only a one-time cost, with little or no recurring cost. The cost to purchase the cables, buses, connectors, etc., is usually reasonable, and once these facilities are purchased and installed they belong to the organization, with no further lease or rental charges. Maintenance costs are typically minimal, so ongoing costs are usually limited to the purchase of new connectors if additional equipment is connected to the LAN.

Many changes are occurring in the United States as the deregulation of the communications industry continues. It is clear that the full-line common carriers (especially AT&T, following its agreement to divest itself of the local Bell System companies) will be allowed to participate more fully than in the past in businesses other than basic communications. On the other hand, more organizations will be encouraged to enter the business of providing network facilities, telephone and terminal devices, and so on. However, to date the exact conditions of further deregulation have not been agreed on within the Congress. The intent of Congress, the courts, and the FCC is clear—to minimize regulation and to introduce a greater degree of competition than at present. However, there is a great deal of concern about how AT&T (and, to a lesser degree, GTE) can be prevented from using a continued monopoly position in certain markets to cross-subsidize competitive services—making it impossible for other organizations without a base of monopoly services to compete fairly. Assuming that this problem is solved adequately, the result ought to be greater variety of services than in the past, with lower overall costs for network facilities.

Network management is also a matter of concern when selecting which network facilities to use. A simple network requires relatively simple network-management capabilities, but even in this case it is necessary to be able to monitor network operation. There are two major purposes in network management: to react to problems or failures in the network and to perform capacity planning. Both hardware tools and software facilities can assist in these tasks. For example, network-monitoring equipment is available from a number of suppliers to assist in diagnosing problems in a network and determining where the difficulty is. Many network-software systems include statistical-accumulation capabilities, so that a network administrator can determine the loading on each part of the network and project trends of use over time.

As networks become more complex, the problems of network management also increase in complexity. It then becomes essential to have a full range of hardware and software monitoring and diagnostic tools, as well as a staff of experienced network administrators to manage the use of those tools. In many organizations, this presents a significant problem; there are no staff members with suitable experience, and it is both difficult and expensive to hire a staff of this type. (Since this type of expertise is in short supply, a seller's market situation prevails.) The use of a PDN or VAN is one possible solution to this dilemma. Typically the supplier of the PDN or VAN provides many network-management functions which the organization would have to undertake if a private network were used. For example, the PDN or VAN supplier provides rerouting if a part of the public network is out of operation, and sometimes if a part of the network is experiencing overload conditions. This type of rerouting occurs automatically, and users of the network are often unaware that it is done. The PDN or VAN supplier may also provide diagnostic messages which the users—perhaps from their terminals—can invoke if there seem to be problems in communicating with another site through the network. The results of the diagnostic can then be reported to the PDN or VAN supplier, who will take appropriate action.

LANs are also simpler to manage than wide-area networks, although they do require some network-management functions. Load management, for example, is necessary on a LAN just as on a wide-area network, because an increased load may lead to delays in response and the need for increased capacity. Diagnostics may also be required, but these are typically much simpler than in wide-area networks, because the network facilities are all located close to one another—and of course LANs are not subject to many of the problems of other networks, such as being affected by thunderstorms, flocks of birds, and similar natural hazards.

Selecting the network facilities is a step-by-step, iterative process, which can be done using the following procedure. This process will indicate which type(s) of facilities are best suited to a particular situation. The selection of specific link speeds and other similar choices will occur as part of the detailed network design which follows completion of the strategic-level design process.

1. *Determine which endpoints (terminals and computers) must be connected and how often data or text will be exchanged.*

All the information necessary for this step ought to be available as a result of the data-collection process and the initial steps of system design. In many systems the requirements for data exchange will form a star pattern, with local and/or remote terminals or work stations exchanging data with a central computer location. In other systems a hierarchical pattern may be formed, with terminals communicating with local information processors, which in turn communicate with a central installation. Still other systems have data-exchange patterns which form a mesh, because many endpoints need to exchange data with one another as well as with one or more information-processing locations. Each of these three patterns is shown in Figure 12-10.

If the pattern of required data exchange forms a mesh, the following points can be considered. If the pattern is either a star or a hierarchy, step 2 can be considered next.

In a mesh situation—which may apply to the entire network or to only some parts of it—the key is to minimize network costs, since the cost of interconnecting many endpoints is typically high. The following choices can be considered:

- LAN

- PDN or VAN

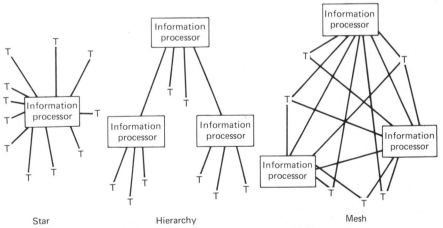

Figure 12-10 Interconnection-requirement patterns.

- Public switched network
- Private meshed network

If the endpoints are close enough together for a LAN to be used, this is the optimum choice because of its low cost. Unfortunately, many systems have endpoints which are so geographically dispersed that the use of a LAN is impractical. However, if even part of the network can be connected using a LAN, this will usually reduce the total cost.

A PDN or VAN or use of the public switched network are the next best choices in the mesh situation. If a PDN or VAN is available, the error qualities will usually be better than on a public switched voice network, but of course a suitable data network may not be available. If both PDN or VAN, and public switched-network facilities are possible choices, cost will be the determining factor. Each network has different rates, and the volume of data transmitted (and in some cases the distances involved) is an important factor in total costs. A study must be made of each network which could be used, to determine what the probable cost would be.

A private meshed network is the final choice for this situation. It is considered the least desirable choice for two reasons: cost and complexity. The initial installation cost of a private meshed network is generally high, and an expert staff of network administrators is required for its operation. As mentioned earlier, there are turnkey network vendors who will install a network of this type on contract, which somewhat reduces the need for an experienced networking staff. There are situations in which a private meshed network will result in lower life-cycle costs, especially if the volume of data to be transmitted is very high. There may also be considerations of security or privacy during data transmission which make the use of a private, unshared network attractive.

If a private meshed network is the best choice, then the exact configuration and link speeds must be determined. A detailed network design is not required at this stage of system design; however, some first cuts must be made in order to estimate network costs. Step 2, which follows, discusses selecting the correct types of facilities based on volume and response-speed requirements.

2. *For each combination of endpoints which must be connected, compare the volume and response-speed requirements to the cost of available network facilities in the locations to be served.*

The best choice in network facilities will depend on the volume of traffic to be exchanged and on the speed of response (and therefore of transmission) which is required. Volume and response requirements are related to the choice of different types of facilities as follows:

- Low volume, with response requirements which are longer than 5 seconds, usually makes a public switched network the best choice.

- High volume, whether or not fast response is required, usually makes private links the best choice. The speed of required response will indicate which capacity (speed) of links are needed.

- Low volume and fast response usually make a packet-switched PDN or VAN the best choice; as discussed earlier, a circuit-switched data network is not usually well suited to this combination.

Of course, whenever the endpoints are geographically located so that connection using a LAN is possible, this ought to be considered before any other type of network facilities are studied. Even if a LAN can be used for only part of the total network, it may reduce costs significantly.

If the public switched network is the best choice, the link speed cannot usually be selected. (AT&T provides high-speed digital switched connections among some cities in the United States, but this is a special case and resembles a circuit-switched PDN more than it does the public switched voice network.) However, in some countries it is possible to obtain modems which determine the speed at which data are transmitted over the network. Speeds up to 19.2K bits per second on voice-grade switched links are possible with appropriate modems. However, it is always necessary to establish the connection before sending data, which reduces the effective speed somewhat. Because the public networks are less expensive than private links for low-volume traffic, and sometimes less expensive than a PDN or VAN, the slower response caused by call establishment may be a valid trade-off to achieve lower cost.

High-volume data transfer between pairs of endpoints usually calls for a dedicated high-speed transmission link. As pointed out earlier, the speed may be achieved by using appropriate modems, if 19.2K bits per second is adequate. For higher speeds, a broadband link is required. Depending on the facilities available and where the endpoints are located, the link may be a land line leased from a common carrier or PTT, a microwave link, or a channel on a communications satellite. As noted earlier, the communications-satellite channel provides a potential additional advantage by making it possible to broadcast the same information from one location to two or more other locations; this is impossible with a land line or microwave link.

Low-to-moderate volume and fast response usually call for a PDN or VAN, if one is available in the required locations. However, it is important to investigate thoroughly the response characteristics of the specific network. Because X.25 meshed networks, including those operated privately, share the links among many types of traffic and many endpoints, it is practically impossible to guarantee that response will be consistent. Some PDN and VAN networks have considerably greater capacity than is normally required, so that response speed will be fast unless an unusual peak load occurs. (This approach was also used in the ARPANET.) Other networks do not have such a high capacity, and response will degrade more often than in other networks. Two things must be considered: first, what statements the network-operating organization makes concerning transit times (which are the network portion of response times) in contractual documents, and second, how important consistently fast response is. In some applications it is not essential to have 1- or 2-second response 100 percent of the time, and variable response speed is acceptable in return for (potentially) lower cost. In other applications high-speed, consistent response is vital; in those cases a PDN or VAN may not be the optimum choice. Point-to-point links, or possibly a private meshed network with excess capacity, may be preferable. However, note that the comments about response consistency in public X.25 networks apply to private meshed networks as well; it is a challenging design problem to configure a network to provide consistent response without spending an inordinate amount of money to overconfigure the link capacity.

3. *If the network includes requirements for fast response and also for the transfer of high volumes of data and/or text, it is usually better to configure separate facilities to meet each of these requirements.*

Regardless of the tentative decisions made in steps 1 and 2, it is important to separately analyze response and volume requirements. It is technically very difficult to mix high-volume traffic and fast-response traffic on the same network links, while adequately meeting both sets of requirements. High-volume traffic tends to saturate links, and if those links are shared with fast-response traffic, response speed may not be fast enough. If the high-volume traffic is broken up into small segments (possibly even into packets) to avoid interference with fast-response traffic, high-volume traffic will probably be delayed and the overhead will be increased because of the control information necessary in each segment—the smaller the segments, the higher the percentage of overhead.

How serious this problem is depends on the specific network and the specific combination of high-volume and fast-response traffic. If the network capacity is higher than might be justified from the volume expected, it will be easier to intersperse the two kinds of traffic. The routing and flow-control algorithms used in the network can also determine how successful integration is. Good routing and flow-control methods allow the total capacity of the network, in all links, to be more effectively allocated among the available types of traffic. The Canadian Datapac X.25 network is said to handle combinations of high-volume and fast-response traffic more successfully than most, because it includes a number of high-speed parallel links between major centers of population and also because it handles load leveling across those links very well. Other networks may be less successful in meeting this challenge.

In general, it is better to separate high-volume and fast-response traffic, especially if the fast-response traffic consists of transactions which serve the main-line operations of the organization. Transactions of this type are typically very important to the organization, and less degradation of response is acceptable than in other situations. If high-volume traffic is moved onto high-speed dedicated links, fast-response traffic can be handled more effectively.

4. *Evaluate the requirements for backup in the network.*

This topic is covered in detail in Chapter 13 because it is one aspect of the integrity and flexibility of the system. Briefly, most networks require detailed planning for how to handle failures, and the greater the importance of the traffic handled, the more urgent backup planning becomes. One of the major advantages of PDNs and VANs is that they provide very useful fallback facilities when used in conjunction with private facilities. A combination of private and PDN or VAN facilities, or possibly a combination of private and public switched-network facilities, is typically a good choice because of the ability to back up one type of facility with the other.

5. *Evaluate the requirements and implications of network management.*

As the earlier discussion points out, both hardware and software tools are required to manage a network. Problems and failures in the network facilities must be pinpointed, so that appropriate repairs can be made. Problems and failures in network software must be isolated and corrected, and of course the tools to perform these tasks are different from those used for facility problems. Overloads, leading to degraded response, must be monitored and, if possible, predicted ahead of time, so that action can be taken to increase capacity before a serious overload exists. In the more detailed phases of network design the hardware, facilities, and software to be used must be carefully evaluated to determine that they provide the necessary tools for the organization's staff of network administrators. Generally speaking, the more complex and widespread the network, the greater the need for comprehensive network-management tools.

It is also important to evaluate the need for skilled personnel to perform network-administrative functions. Although this is still early in the system analysis and design phase, it is by no means too early to consider staffing. The more complex the planned network, the greater the need for experienced personnel—or for personnel with suitable background in networking hardware and/or software who can be trained as network administrators. Those people ought to be selected before the final, detailed design of the system is begun, so that they can participate in that process. If the people who will have to operate and administer the network are involved in its design, their requirements are most likely to be met, leading to much improved operability for the entire system.

6. *Estimate the cost of the most suitable network facilities, and determine if this cost is within budget constraints.*

Because the selection of network facilities is tentative, a detailed evaluation of network cost will not be possible. However, it ought to be possible to estimate the total monthly cost and one-time costs for items such as purchased cables, etc. Typically, the one-time costs, unless these include the preparation of room(s) to house network-control hardware, can be ignored, since they are small in relationship to the monthly recurring costs. Depending on how the budget process operates, the monthly costs may be the only important number or it may be necessary to estimate life-cycle costs, including facilities, staffing, and any other pertinent items.

To prepare a simple, first-cut cost estimate, compute the monthly rate of each type of facility to be used. If a network consultant is involved in the strategic-level design, the consulting contract ought to call for preparation of this cost estimate. If the local staff of designers is solely responsible and does not have the background to estimate the network costs, assistance can be obtained from AT&T, the local telephone supplier, the PTT, and/or the operator of the VAN or PDN being considered. While all such cost estimates must be treated carefully because of their preliminary nature, they can serve to indicate whether the design process can continue or if a significant reevaluation of the design to this point is required.

If the estimate of network costs is within budget constraints, then the system-design process can continue. If it is not, then steps must be taken to minimize costs in some way. Sometimes this effort can be put off until the more detailed phase of network design. If the estimate is within 10 to 15 percent of budget, this may be the best move. If the estimate is clearly much higher than can be tolerated, steps must be taken during strategic-level design to lower the cost. Some possible actions are as follows.

• *Make use of cost-minimization techniques common in networking.* These techniques are outside the scope of this volume; sources of this information are included in the Bibliography. To summarize briefly, some basic techniques which can be helpful are: (1) use multipoint links instead of point-to-point links attached to each endpoint; (2) use switched connections instead of dedicated links; (3) use lower-speed links and/or lower-speed modems; (4) reconfigure the network links to lower the distance traversed; (5) use multiplexing or concentration to combine multiple endpoints onto a single physical link. All these techniques have been in use for many years and may significantly lower costs in some cases.

• *Consider the distribution, or further distribution, of functions.* This is also becoming a reasonably common technique for minimizing network costs. If processing functions can be moved close enough to the terminal users, the endpoint devices can be connected to the processor(s) by LAN facilities instead of a wide-area network. There may still be requirements for access to one or more central locations on an occasional basis, but the volume and frequency of transmission are typically lower and therefore the cost is less.

If this approach is chosen, it involves reconsidering the decisions made according to the guidelines in Chapters 10 and 11. In many cases further distribution will prove feasible; however, it is important to analyze the costs which may result from this decision. Moving functions from a central location to remote locations may not lower the cost of the central processing system(s) enough to cover the cost of the satellite processors. These additional costs must be subtracted from the potential savings in network costs.

In other cases the distribution of functions and data may be impractical; for example, security may make it essential to keep databases centralized. The distribution of functions to lower network costs might result in the need to build additional computer centers—a capital investment the organization might be unwilling to make. This possibility therefore requires careful study before it is selected as the solution to apparently excessive network costs.

• *Determine if all prospective users of the system need terminals.* This possibility ought to be considered only as a last resort; however, in a number of cases a decision is made to provide terminal devices only to selected users rather than to all of them. If, for example, some users are located in very remote areas and the density of the user population in those areas is low, it may be impractically expensive to provide terminals and on-line access to all those users. Limiting terminal use to clusters of users in areas reasonably close to the central processing location(s) may lower costs significantly. Other methods of input and output must then be provided for those users who will not have terminal access to the information system. Methods such as voice telephone, facsimile, or even the mail may be suitable, depending on the specific application situation. This is clearly not a desirable solution, as it may limit the new system's expected payback in increased efficiency. However, practically speaking, it may be necessary to make this type of decision to ensure the financial feasibility of the new application.

After carrying out the above six steps, the result will be a high-level design of the communications network for the new system. This design will indicate the type(s) of facilities to be used, which users and processing locations will be linked by those facilities, what facilities and staffing for network management will be provided, what basic structure (star, hierarchy, meshed) will be used in the network, and what the estimated cost will be.

EVOLVING FACILITIES TO BE CONSIDERED

Some new network-based services are becoming available and are worth consideration during high-level network design. These are described as an appendage to this chapter, since the facilities discussed are not widely available yet. The system designer must be aware of these evolving capabilities and become familiar with the situation in the geographical areas of interest. When appropriate, these new offerings can then be integrated into the planning for new networks.

ELECTRONIC-MAIL SERVICES

Electronic-mail services are becoming available in a number of forms. In the United States some carriers offer an electronic-mail option as part of a value-added service. GTE-Telenet, for example, provides a service called Telemail, which allows a subscriber to the network to send messages from a terminal to another terminal or computer attached to the Telenet value-added network. The major difference between electronic mail and interactive message exchange is that the addressed party need not be on-line when an electronic-mail message is sent. (Interactive messages can be exchanged only when both parties are on-line simultaneously.) The network provides for the storage of each message, with automatic delivery whenever the recipient is available. Other PDN and VAN carriers, including Tymnet, provide similar services. The optional features vary and may include receipted mail, in which the sender is notified when the recipient has received the message. Also, the recipient may be notified of any waiting mail whenever he or she connects to the network and be given the option of selecting any, all, or none of the messages for immediate delivery.

If electronic mail is a requirement in the system, and the endpoints which will use this facility could be connected to a VAN or PDN that offers this service, it may be the most efficient and cost-effective method. No implementation effort is required, although the provider of the service may require that the subscriber's administrative staff supply parameters defining which endpoints will use the service, what the delivery rules are, and so on.

Electronic mail is only the first of the facilities, in addition to data transmission, which are provided by value-added carriers and the PTT suppliers of PDNs. It can be expected that these organizations will move further into the realm of data processing to provide innovative services. AT&T, for example, in its original description of the Advanced Communications Services (ACS) value-added network, described features such as data entry and data transformation (code conversion, reformatting) as network functions. U.S. government regulation of the common carriers has prohibited many technically feasible offerings, but the trend toward deregulation seems to be gradually eliminating those restrictions. In PDNs, government policy will determine whether services of this type are offered; typically each PTT will perceive these offerings as a good business opportunity, just as AT&T does. It is therefore useful to continually monitor the capabilities which are being introduced in these networks and evaluate their relevance to information-system requirements.

VIEWDATA

Viewdata is used here as a generic name for new capabilities which are based on the delivery of computer-based services via a television set or video-display terminal. The ultimate aim of viewdata is to provide data access and services to people by TV in their homes. However, at present there are many more business, government, and similar applications of viewdata than there are uses in the home. It is therefore a possibility worth considering in at least some computer-based information systems.

Viewdata can be provided in two forms, one called *teletext*, in which the user can access data but cannot provide input, and the other called *videotex*, in which the user can interact with the system. If the terminal used is a TV set, it must be modified with additional electronics and an expanded remote-control keypad which allows for the selection of data to be viewed and, in the videotex version, for data input.

Data accessed by viewdata is stored in a computer database, but in the form of screens (sometimes called pages) of data, each of which is a display for the TV or VDT screen. Viewdata uses fixed-frame TV, meaning that each display is shown on the screen independently, without motion. Fixed-frame TV can be transmitted over regular voice-telephone lines, which is one method of delivering the data to the TV or terminal. Full-motion video, in contrast, requires a much wider bandwidth, and therefore must use broadband communications links—which are too expensive for practical use in most viewdata applications today. Other methods of linking the TV or VDT to the database are TV broadcast and cable TV, each of which provides enough bandwidth for motion; however, motion is not yet used in viewdata. Broadcast TV supports only the teletext, or output-only, form of viewdata, while cable TV and telephone lines can support either teletext or interactive videotext.

There are a number of operational and experimental viewdata systems in place. In the United Kingdom the Prestel system was one of the earliest viewdata systems. Prestel uses the public telephone lines and allows users to access a variety of databases for information and amusement. An example of a possible viewdata screen in a system such as Prestel was included in Chapter 9 as Figure 9-7. These systems are menu-driven and use the same methods as in menu-based VDT interfaces (described in Chapter 9).

Other operational systems include Antiope in France, Telidon in Canada, and several experimental systems in the United States. France has embarked on a very large viewdata-type project, in which each of 30 million telephone subscribers will be given, free of charge, a display terminal with which to access the PTT's telephone directories. When all the terminals have been installed, no further telephone directories will be printed and directory-assistance services will be minimized. It has been estimated that the cost of the 30 million terminals will be rapidly repaid through the savings from printing and directory assistance. In addition, the terminal volume resulting from this project allows the participating French terminal vendors to attain very great economies of scale, making them extremely competitive in terminal sales in other countries. (One large sale in the United States in 1981 resulted from this competitive advantage.) The terminal used for this purpose will also be adaptable to limited viewdata use—limited, because of the small screen size.

In the United States one of the Bell System telephone companies proposed to implement a directory-assistance system similar to the French system, but for yellow pages (advertising) only. However, many organizations objected to this experiment on the grounds of an unfair monopoly situation and the interference of AT&T with newspaper advertising, radio and TV advertising, and so on. As a result, the experiment was canceled. It is, however, only a matter of time before such services are begun in some form, in line with the present deregulatory sentiment in the United States.

Because there is no viewdata service sponsored by AT&T (due to regulatory restrictions), experiments in the United States have been much more fragmented than in many other countries. These experiments have been, and are being, carried out by a number of organizations but tend to center around two possible applications: home shopping and home banking.

In home-shopping services, the shopper can browse through a catalog of available items using the TV set. Some experiments simply use this as a means of advertising and expect the shopper to either visit a retail store or make an order by telephone after selecting the goods to be purchased. Other experiments are designed to allow the shopper to select the goods using the viewdata system and either charge the purchase to a credit card or initiate a funds-transfer transaction with a participating bank.

Home-banking experiments have typically used display terminals rather than the TV set, since there is no need to display pictures as in home shopping. The services offered in home banking include checkbook management, automatic bill payment, and customized services specific to the bank. Some customers are reported to be very pleased with these services. However, the cost of the terminal has so far been a deterrent to wide customer acceptance. If the bank provides the terminals free of charge, the service is too costly for the bank; if the bank charges the customer for the terminal, the cost must be low enough to be offset by the increased convenience of the services. So far, an acceptable terminal device at an acceptable price has not been demonstrated. However, as in the case of the French PTT directory system, this is largely a matter of volume.

Some of the systems discussed are public or semipublic viewdata services, open to many subscribers and with shared access to a common database of data or services. It is also possible to implement a private viewdata service, as has been done in some experimental systems. In these private systems, the subscribers are linked by dial or dedicated telephone lines to an information-processing system owned by the sponsoring organization. Access can be through either TV set or display terminal. An example of an application in this environment might be order entry. The subscriber, who might be either a customer of the organization or a salesperson employed by the organization, calls up a menu showing which transactions can be carried out. If the desired transaction is an order, the next menu asks which category of items is to be ordered. That selection leads to the display of the first screen listing items in that category. When the desired items are displayed, the user keys in the required quantity and probably the required date of receipt. Since the connection is directly to the computer, on-line processing of the transaction is possible, and the computer can display the result of the order processing immediately. The customer or salesperson will therefore know at once whether the order has been accepted or back ordered, and if accepted, when the shipment is scheduled. This sequence is in fact almost identical with more typical transaction systems, except that TV sets are used instead of terminals.

The possibilities for the use of viewdata in business, government, education, and other organizations are almost unlimited. Since, effectively, any location can be linked with any other location, assuming that each is served by a telephone network or other shared network such as cable TV, it would be possible for many people to work at home and yet communicate with others in the same organization. Customers might directly access the computers of their grocers, bankers, lawyers, doctors, and others. The use of the TV set—ubiquitous in many countries— opens up new possibilities that would remain uneconomical if terminals were to be used instead. However, the full impact of viewdata will not be felt for a considerable period of time and cannot even be predicted with any accuracy today. Even before the "wired society" becomes prevalent, however, each system designer must become familiar with this technology and track its evolution, so that it can be applied where and when appropriate to new computer-based information systems.

13

DESIGNING FOR INTEGRITY AND FLEXIBILITY

This chapter covers several important aspects of system design which must be considered in coordination with the design of the processing functions, the database structure, and the network facilities. Ensuring that the system can achieve the required level of integrity and provide the necessary degree of flexibility affects all three of the areas discussed in Chapters 10, 11, and 12. Unfortunately, integrity and—especially—flexibility are often considered peripheral to the main system-design decisions and given less than adequate attention. This can lead to an unsatisfactory system design and may cause a redesign and reimplementation of portions of the system after implementation. Devoting a chapter to these aspects of design reflects their importance and emphasizes the attention which must be given to them throughout the design process.

It may appear that integrity and flexibility are relatively unrelated topics, so that discussing both in a single chapter may seem unusual. However, on closer inspection it is apparent that there are many design techniques which apply to both integrity and flexibility. Other aspects of integrity and flexibility are unique and must be treated independently. This chapter covers both the common design techniques and those which apply to only one of these topics. Before discussing how to design for integrity and flexibility, it is important to define what each of these terms means within the context of system analysis and design.

INTEGRITY

Integrity is used here as a general heading to cover the following closely related aspects of computer-based information systems:

- Correctness
- Availability and survivability
- Security
- Privacy
- Auditability

Although the above list covers a wide range of subjects, all are related to the concept of integrity, also defined (in Webster's) as soundness. A system which has all the above characteristics has a high level of integrity.

- *Correctness* is the degree of accuracy of data managed, processed, or transmitted by the information system.

- *Availability and survivability* are ways to measure whether the system, or an adequate subset of the system, will be usable whenever the users require access to its facilities. Availability is used here in the general sense of reliability, while survivability refers to the ability of the system to continue operation even though problems occur or some components fail.

- *Security*, which is related to integrity because a base of correctness and reliability is required to ensure security, covers all aspects of the protection of the data and facilities of the system—except for the protection of personal data, which is included under *privacy*.

- *Auditability* covers system characteristics which make it possible to formally examine the system to determine the correctness of its operation and the results of that operation.

FLEXIBILITY

Flexibility, the other topic covered in this chapter, includes the following system characteristics:

- The ability to modify functions, protocols, or interfaces provided by the system or to add new functions, protocols, or interfaces

- The ability to expand the capacity of the system (or possibly to decrease the capacity)

The flexibility to modify or add can apply to any type of system functions, to protocols for communication between any pair of system elements, to external user interfaces or internal interprogram interfaces, database content or structure, and/or network facilities—including links, network processors, terminals, and other devices such as multiplexors.

The flexibility to expand the system's capacity may apply to the number of users handled, the number of terminals connected (simultaneously and/or in total), the number of transactions processed, the number of jobs processed, and/or the number of ad hoc queries or reports processed. In some cases system capacity may decrease rather than expand. This may result from moving functions and/or users from one information-processing location to another in a distributed system, or possibly from an overall reduction in the organization's workload. Decreases in capacity are usually technically easy to handle, although their impact on system cost-effectiveness may be more difficult to manage.

GROUND RULES

Ground rules which apply both to designing for integrity and to designing for flexibility include the following. Design techniques unique to each of these areas are discussed later in this chapter.

1. *Design for the maximum feasible modularity. Use the piece-at-a-time approach (see Chapters 1 and 3) rather than the grand-design approach throughout system design and implementation.*

   It is perhaps a cliché that modularity improves flexibility, but clichés are usually based on experience. Modular systems *are* more easily changed, expanded, or contracted than monolithic systems. It may be less apparent that modularity improves integrity. However, a modular system is more readily understood and is less likely than a more integrated system to have hidden relationships among its elements which eventually cause errors or failures. Modularity is therefore one of the most basic ground rules of good design.

2. *Design for simplicity and avoid complexity.*

   This rule is closely related to the modularity rule, and is similarly a basic tenet of good design and programming. Complexity in design or implementation makes a program or system difficult to understand, and systems which cannot be understood cannot easily be changed. They are also very apt to have complex interrelationships which cause incorrect operation or failure when specific sets of circumstances arise. Simplicity, like modularity, therefore contributes to both integrity and flexibility.

3. *Create a set of design and implementation standards within the organization, and observe them rigorously.*

This rule addresses internal standards, which are set by the organization to meet its specific needs. However, in many cases the internal standards will simply reference external standards. This is the best policy in each case for which an external standard exists, since the use of industry standards makes the acquisition of hardware, software, and communications facilities much easier than when unique internal standards are defined. The areas which require attention for standardization include the following, which apply to all organizations; some groups may determine that additional standards unique to local requirements are needed.

- *Programming-language standards* are essential, as the ability for one implementer to understand another's program (module, etc.) depends on the use of the same language or language subset. Industry standards exist for languages such as COBOL and Fortran and are being defined for new languages such as Ada. The internal standards which reference external standards must define explicitly which subset of the total language is to be used; language standards typically define various levels and/or options, so simply referencing a standard is inadequate.

- *Protocols*, especially networking and distributed-system protocols, must be standardized. As Chapter 12 points out, there is a very high level of activity in standardization today, and many products are available which comply with the newly defined standard protocols. It may in some cases be necessary to set long-range goals for evolution to standardization on new protocols such as HDLC and X.25. Each new implementation must then be monitored to determine if the new protocols can be used—or if not, the use of older protocols must be justified.

- *Interfaces*, especially terminal-user interfaces, must be standardized. As Chapter 9 points out, the design of user interfaces is one of the most important aspects of a computer-based information system—and good interface design is crucial to success. A set of guidelines for the design of interfaces, based on successful experience within (or outside) the organization, can ensure that each system designer will make use of others' experience. While a "user-friendly" approach to interface design is essential, it is equally important to ensure that user interfaces are consistent; i.e., that each designer does not use a different approach. While each such approach might be well designed, if one user were required to use different interfaces for different tasks, confusion would surely result. Standards can prevent this confusion, allowing users to see a compatible interface for each similar function.

· *Internal interfaces* between modules, programs, and subsystems must also be standardized. This is an essential element of rule 1, *designing for modularity*. Well-defined interfaces allow each module, program, or subsystem to be designed and implemented independently of the elements with which it communicates. In addition, standardization allows interfaces to be changed without affecting the components which use those interfaces (if care is taken in component design), and each component can be changed independently as long as the interfaces remain stable.

4. *Document everything*

This may seem to be a very broad rule, and it is possible to create too much documentation. In practice, however, that problem seldom arises; in fact, most systems suffer from a severe lack of documentation. A system which is not documented cannot be changed and is therefore inflexible. A system without documentation cannot be understood and is therefore unstable and cannot possess a high degree of integrity. Self-documenting flowcharts and programs provide one solution to this problem, and there are software systems available which can assist in the creation of these charts and programs. Many installations have turned to the use of documentation specialists (who are technical writers with programming or design background), whose role is to provide the necessary documentation as the system design and programming procede. This is a very good solution to the problem and is probably more practical than expecting all designers and programmers to become facile writers.

The above four rules apply both to designing for integrity and to designing for flexibility. Other techniques which apply to only one of these areas are described in the remainder of this chapter.

DESIGNING FOR INTEGRITY

The components of integrity were defined earlier: correctness, availability and survivability, security, privacy, and auditability. Well-chosen design and implementation techniques can help to ensure the degree of integrity needed in each area. High integrity becomes increasingly important as main-line functions are directly supported by computer-based systems. It is therefore almost impossible to place too much emphasis on system integrity.

CORRECTNESS

Correctness is perhaps the most basic component of integrity. Correctness applies to data being entered into the system, to the content of the system database(s), and to the data transmitted.

Data-Entry Integrity   Correctness must begin with the data elements provided as input to the system, because if these are incorrect all subsequent processing will be incorrect. There are three rules which can help to ensure that data entered into the system are correct.

1.   *Minimize the amount of keying required in data entry.*

Although some people who enter data are skilled at keying, many are not. As data entry becomes more often a by-product of the main-line business or government processing rather than a separate function, terminal users will be less likely than in the past to have typing experience. Minimizing the number of keystrokes required is a very simple way to minimize the possibility of error. (Even when data-entry personnel are skilled typists, statistically there will be fewer errors if there are fewer keystrokes.) As discussed in Chapter 9, there are a number of ways to minimize keystrokes, including the use of menu-driven systems—which is probably the most common method of data input. Techniques such as the use of badges, magnetic cards, light pens, or touch-sensitive screens can also minimize keying.

2.   *Design the data-entry interfaces specifically to meet the needs of each group of users and the environment in which they work.*

This rule may seem to contradict the earlier rule regarding the standardization of user interfaces; however, it is possible for interfaces to be both standard and custom-tailored. Standards must define the types of interfaces to be used; for example, a standard might specify the use of menu-driven interfaces rather than free-form data entry. Standards may also define the specific vocabularies to be used for different functions and/or different classes of users. Within those standard guidelines, it is possible to customize interfaces for each specific purpose. For example, if a menu-driven approach is used, order-entry personnel will need a menu which allows them to enter data in a logical sequence—based on how customers are likely to provide the data and/or on how orders are recorded on existing forms. The specific sequence and format of the menu, and the terminology used in building the fixed portion of the screen content, will be designed for those specific users. These interfaces will, however, have overall consistency with other data-entry interfaces, such as those used by financial personnel.

The rationale for this rule and the reason it is included under correctness is that interfaces which are well designed and easy to use are helpful to the person who must enter the input. In contrast, interfaces which are poorly designed for a specific user environment make data entry more difficult and errors more probable.

3. *Define extensive data-editing routines to be activated during input.*

This rule is followed in almost every system today but is still worth stating. Detecting errors at the point of data entry is essential, as later detection involves more complex methods for correction. Every essential data element *must* be validated at the time of input. Using intelligent terminal devices or controllers to perform these functions improves system efficiency enormously.

Database Integrity   In addition to data input, the *content* of the system's database(s) must be correct. Emphasis must be placed not only on new data entering the database but on each update to the content. Database updates which take place on-line must be validated when they occur, as the information necessary for correction may be difficult to obtain later. Some ground rules to assure a high degree of accuracy in database content follow.

1. *Define extensive data-validation routines which apply to each update and which analyze both the data used for the update and the results of the update process.*

Some DBMS-software systems have built-in data-validation routines, but if these are not available in the DBMS being used they must be created locally. Each data element used to update the database must be validated to ensure that it is within reasonable limits. The result of each update must similarly be validated for reasonableness. Any suspect update must be rejected, and displayed to the terminal user who initiated the update and/or to a database administrator for analysis and correction. The user must not be allowed to override validation checks except with the approval of a database administrator.

2. *Do not allow access to, or update of, a database except by standard DBMS software.*

Programmers are often tempted to bypass the DBMS software and access the database content directly. This may appear attractive—to improve performance by avoiding "extra" layers of software with their attendant overhead. However, this shortcut is very shortsighted and can result in damage to the content of the database in the form of incorrect data elements and/or an invalid structure.

One of the standards mentioned earlier in this chapter ought to define how database access is allowed. Design reviews and program reviews must ensure that those standards are not bypassed. Some new systems, such as IBM's System/38, are organized so that microcode protection ensures that standard database-control software cannot be bypassed. It is likely that other systems will be implemented using this approach in the future, because database integrity is too important to jeopardize through poor programming practices.

3. *Ensure that the DBMS to be used has an adequate set of concurrency-control mechanisms.*

Database updates which occur on-line can interfere with one another; if the DBMS software (or the operating-system software) does not protect against this potential problem, the conflicting updates may damage the content of the database. Generally speaking, it is reasonably safe to assume that DBMS software from a major hardware or software vendor has adequate concurrency control, but if any DBMS-related software is obtained from other sources, this aspect must be investigated.

Transmission Integrity   The final aspect of correctness is how to ensure accurate data transmission.

1. *Choose a contemporary, bit-oriented protocol such as HDLC or SDLC, which has the most adequate error-detection scheme generally available.*

Older protocols do not find as high a proportion of transmission problems as the newer protocols, which use cyclic-redundancy checking to detect errors. If an older protocol is used, which is less complete in its error-detection capability, it may be necessary to include validation routines at the receiving location to further check the data before their use in processing.

Whenever design techniques to ensure data correctness are being evaluated, it is important to understand the level of correctness required by the application. This topic is discussed in Chapter 5. To restate the major point of that discussion, some applications—especially those which handle documents and text instead of data—do not need as high a level of correctness as others. It is pointless to expend additional computer (and/or personnel) resources to obtain a higher level of correctness than is necessary, as this increases the total cost of the system.

AVAILABILITY AND SURVIVABILITY

Availability and survivability are extremely important aspects of system integrity. There are a number of ground rules which can be observed during design to ensure that the system is adequately reliable. These ground rules cover methods to provide high availability for information processors, for the system database(s), for network facilities (links, network processors, etc.), and for terminals. The relevant guidelines are listed for each of those categories.

Information-Processing Availability

1. *Provide automatic restart and recovery capabilities for all the system's information processors—host and satellite (if used).*

   The systems being implemented today cannot afford to cease operation while manual recovery processes are underway. The system software must therefore provide for automatic restart after any type of failure (whether caused by hardware, software, or facilities such as electrical power), with recovery of all in-process operations. Hardware features such as a *programmable read-only memory* (PROM) may be required in some cases to support automatic restart. In certain cases manual intervention will be required, but this ought to be limited to functions such as the reentry of partially completed interactive transactions, the mounting of necessary backup reels of tape, and so on. Systems which do not support automatic restart and recovery are unsuitable for most new computer-based applications.

2. *Ensure that satellite information processors and intelligent terminal devices can operate unattended, including the ability to carry out fully automatic restart and recovery.*

   Programmable devices installed in locations other than computer centers must typically be able to operate without a computer operator. It is uneconomical to station an operator at each remote location, and it may also be infeasible to train remote personnel in complex computer-operating procedures. (It is acceptable to train personnel to turn power on, mount paper in printers, and other similar tasks.) Mini- or micro-based remote equipment may be less likely than general-purpose computers to have fully automatic restart capabilities, and yet it is often more important on those systems than on large computers. In a computer-center environment, it is reasonable for the operators to assist in recovery operations, such as mounting magnetic tapes, disk packs, and so on. Satellite processors, in contrast, must be able to restart completely independently and/or accept program and data files down-line from a central location to accomplish restart. If the software (and hardware) facilities required to accomplish this are not provided with the system to be used, they must be created in-house. Of course it is better to acquire remote systems which already have these features, as it may be expensive and time-consuming to implement them.

3.   *Consider the use of hardware and software which allow for graceful degradation when problems occur.*

The use of multiple components, such as processors, memory modules, etc., in a single system allows that system to gracefully degrade if a component fails or is malfunctioning. Tandem Computers Incorporated provides high-uptime systems explicitly designed for on-line use and consisting of sets of dual hardware and software components. Any single failure, and even some combination failures, can occur without causing a system failure. Another company, Stratus, has more recently introduced redundant-component systems that provide dual components of which only one does useful work at any time. Failure of a component therefore does not cause the system to fail and does not even degrade the performance of the system. This approach has only recently become economically feasible, but as the cost of hardware logic decreases rapidly, it becomes attractive to use backup logic to increase availability.

Other systems, although perhaps not specifically designed as redundant or high-uptime systems, provide similar capabilities. For example, the large-scale computers produced by Burroughs, Honeywell, and Univac all allow the configuration of multiple processors operating under a single copy of the operating system, as well as the configuration of multiple main-system components such as memory modules, I/O controllers, and so on. Failure of any of those elements for which one or more other copies exists causes the system performance to degrade but does not cause system failure. In high-uptime applications this type of configuration is more appropriate than a single processor—which must typically be backed up by another single-processor system.

Database Availability

1.   *Ensure that the DBMS software to be used includes deadlock detection and resolution.*

All DBMS systems have the potential to develop a *deadlock* (also called *deadly embrace* or *mutual interlock*) if more than one program (or module) is allowed to update the database concurrently (or simultaneously, in a multiprocessor system). In a deadlock situation, one program has locked a record and is waiting to obtain access to another record locked by another program. The second program, meanwhile, is waiting to obtain access to the record locked by the first program. (The same situation exists if the locking is at the field (data element) level, although there is usually less likelihood of encountering a deadlock.) Unless the software has built-in mechanisms to detect this situation and force a break, the deadlock may persist indefinitely. The remedy is to force the termination and rollback of one of the deadlocked programs so that the other can continue. After a pause to avoid re-creating the deadlock, the terminated program is reinitiated.

In a distributed database, deadlock can be more difficult to detect and may cause more complex problems. While the distributed database and application design may make deadlock an unlikely occurrence, this can seldom be guaranteed. Deadlock in this situation can be avoided by allowing a program to have outstanding update requests in only one location at any time. If each program is updating at only one location, it can cause a deadlock at that location but it cannot cause a multisystem deadlock. The former is reasonably easy to detect and correct, with an acceptable level of overhead. The latter requires a considerable amount of overhead—in some cases, so much that it is impractical.

2. *Ensure that the DBMS software provides for database rollback, in case of application failure, by writing "before images" of the database content and automatically reapplying the "befores" in case of failure.*

This capability is almost universal in general-purpose DBMS packages but is not necessarily provided in a mini-based or micro-based DBMS. In some cases it may be acceptable to recover a satellite processor's database by transmitting a new copy from a host location by down-line load, but this is usually time-consuming. The ability to recover the database immediately, using before images, is far preferable.

3. *Investigate the capabilities provided to dump the contents of the database, both continually, in the form of after images during update operations, and periodically—and to restore the database content through roll-forward in case of serious problems or hardware failure.*

This is one of the most severe problem areas in system design as databases become larger and larger. Originally, many databases were copied completely to magnetic tape each night. Even today, some databases are dumped daily, but it takes a long time to carry out this type of dump for a database with hundreds of billions of bytes. There are several techniques which can minimize this problem somewhat, although none which can eliminate it. One possible method is to dump the database continuously, on a rotating basis, so that over time a completely new dump copy is created. Since it is usually necessary to continue updating the database during the dump process, the after images must be coordinated with the incremental dump to produce a true picture of database status. Another method is to dump a segment of the database each day, when no updates are in process—at least to that portion of the data. Like the rotating dump, this eventually produces a complete database image, and the process can then begin again. High-density streamer magnetic tapes increase the speed of dumping, as they minimize the difference in data-transfer speed between the disk devices and the tape and can be used to shorten the length of time needed for dumping.

In some cases a distributed database is used as a backup copy of the data, with geographical separation for further protection. This method, however, is difficult to manage, especially if both copies of the data are updated on-line. It is possible to use a remote data copy in lieu of a dump, but very careful design is necessary.

One pitfall which some organizations overlook is that restoring the database from dumps and after images may not only be time-consuming but involve very complicated procedures. This was not the case when a full copy of the database was made each night; in that case, reloading was very straightforward. If incremental or rotating dumps are used, however, the dump copies must be very carefully and accurately integrated with after images of updates which occur during or between dump cycles. This may sound simple in theory, but in practice—when confronted with a damaged database and a large number of reels of tape containing dump and after information—it becomes dauntingly complex. It is important to have explicit procedures for database restoration—either a full restore in case of major problems or a partial restore for less severe difficulties. Also, as in every disaster-backup plan, it is important to conduct a dry run of the procedures periodically to ensure that they work as designed and accomplish the desired result.

4. *Consider database duplication if other methods of database protection are inadequate.*

Although the duplication of database information is expensive and presents potential problems of synchronization between the copies, it may be worth consideration. Especially in a distributed database, where copies are physically remote and therefore presumably cannot be affected by the same problems of the physical environment (e.g., power failures), this may be necessary. The ideal situation is a replicated database in which only one copy is updated on-line. The other copy therefore represents the status of the active copy at a known previous point in time. If after images are retained during update, the copy can be brought up to any desired point by applying the afters.

This approach is not usually feasible if the database is very large, because of the time and expense required to transmit the restored copy back to the home location of the database. If, however, a communications-satellite channel or microwave channel is available, the time and cost may be acceptable. A variation on this approach duplicates only essential portions of the database on-line, either at the same location or another location, to minimize the time and expense of recovery.

5. *Consider how to protect remote database segments, whether copies or partitions.*

If a distributed structure is chosen for the system database, special attention to the protection of remote database segments is required if they are placed in locations which are not typical computer rooms. Often database partitions or copies are attached to satellite information processors in offices, stores, factories, warehouses, and similar locations. It is not usually possible to protect those segments by dumping data to magnetic tape.

If the remote segment is simply a replicated copy which is used only for inquiry, it can be protected by backup from the master copy. This may involve a delay if serious damage to the remote segment occurs, but, on the other hand, damage is less likely if no updates are performed. (Only hardware problems are a threat in this situation.) If the remote segment is updated, before images must be created, just as in a centralized database. These can be stored on disk (or possibly diskette) and used for recovery if application-program errors occur. After images can also be written to disk or diskette. It would also be theoretically possible to protect the database segment by a periodic disk-to-disk dump, but this is expensive and seldom practical. Instead, the master copy of the segment is used as a remote backup for restoration in case of serious problems. If the master segment is not updated on-line, while the remote segment is, then restoration from the master plus the after images is straightforward. If both the master and remote copy are updated simultaneously (and the same data elements are updated independently at each location), a very careful analysis of how to restore the copy is required. In fact, this situation ought to be avoided, because the recovery procedures are so complicated that it is very difficult to ensure that restoration will be successful.

If the remote database segment is not a copy but a database partition, it must be protected independently of any other copy. It may, in fact, be necessary to use the typical dump process, even if that requires extra disk space and/or the use of magnetic-tape drives at the remote location(s). If a very-high-speed link to a central location is available—such as a microwave channel or communications-satellite channel—the database segment can be dumped over the link to that location. Of course, this is not always practical because of the expense and time involved, but it has the added advantage of eliminating the need for magnetic-tape equipment and also removes the backup copy physically from the database-segment location.

It can readily be appreciated that the protection of a distributed database, in terms of availability and survivability, can be quite complex. The system designer must make every effort to reduce this complexity to the lowest possible level, as complex procedures usually cause errors, resulting in a loss of database integrity. The simplest methods and procedures which can be devised to accomplish the stated goals are always the best.

Network Availability

1. *Ensure that the communications links, processors, and other equipment, such as multiplexors and modems, have adequate diagnostic capabilities, and/or that additional diagnostic equipment is planned and installed.*

One of the most difficult problems in large networks is determining the cause of failures or incorrect communication. If, for example, a remote terminal device appears to be malfunctioning, it may be very hard to determine if the problem is in the terminal device, its modem, the link connecting it to the next element in the network, that element, and so on—up to and including each software element of the information processor serving the terminal.

Many devices have built-in or optional diagnostic capabilities; for example, many modems sold in the United States have optional diagnostic features. Some of these can be activated remotely, by program control, while in others the modem must be manually switched into diagnostic mode. Many terminals now include self-test routines which are executed automatically each time power is turned on and which can be initiated at other times by the user. Some organizations may consider diagnostic options to be an unnecessary extra expense—however, consider the alternative. In a widespread network, problems must be pinpointed; this can be done either by using remotely initiated diagnostics or by sending technical personnel to the suspected source(s) of the problem. Over time, the latter is prohibitively expensive, even if the expert technical staff to perform those functions can be obtained. An initial investment in diagnostic tools leads to a large saving over the life cycle of a typical information system. (There may be some exceptions; for example, when the entire information system is geographically limited, perhaps within a major metropolitan area. In that case, diagnostic tools may still aid in problem resolution, but the cost trade-offs will be different from those in a more widely dispersed system.)

During system design it is essential to consider the different problems which can occur, how each type can be pinpointed, and what action will be taken to resolve each problem. One approach is the use of brainstorming sessions at which participants attempt to define all the possible problems, diagnostic procedures, and methods of correction. The participation of several people, with different backgrounds in hardware, software and networking, can provide a wide range of possible approaches in a relatively short time.

2. *Consider how to reroute network traffic if links, network processors, or other elements fail.*

If network traffic is essential to continued operation of the application—as in a reservation system, for example—traffic must be rerouted if a particular path through the network fails or causes a high number of errors. If a PDN or VAN will be used, this problem can be ignored, as the supplier of the network will provide the necessary rerouting (although it is a good idea to be sure that this is true for the specific network to be used.) If a private meshed network will be used, the alternate paths which are inherent in the network structure provide readily for rerouting. Of course, the network-routing algorithms must ensure that traffic takes advantage of the network's alternate paths.

In other types of networks, specific planning for rerouting is needed. In a private point-to-point network, backup links can be configured and used at all times, with fallback to a single link in case of failure. The public switched network can also be used as backup to the private links. The choice of one of these alternatives must be based on cost and the amount of delay which can be accepted during switchover. If minimal delay is required, it is better to configure backup private links so that switchover can be instantaneous. (This has the added advantage of providing greater capacity when both links are operational.) The cost of this approach is, however, usually higher than when the switched network is used for backup. It may also be possible to use a PDN or VAN for backup, and in that case arrangements must be made to have the necessary network connections installed and ready for switchover in case of failure.

3. *Plan for the necessary level of backup for terminal devices.*

Failure of a terminal device means that the user of that device cannot access the information system, so the design must provide for alternate ways to serve the user. Although terminal devices do not fail very often, when one does, the repair process is typically lengthy. Either the device must be taken to the vendor's repair location and exchanged for another device, or a repair technician must visit the site where the terminal is located and repair or replace it. (Terminals are becoming so inexpensive that replacement may be preferable to on-site repair.) It must be assumed, therefore, that a failure means a minimum of several hours during which the user does not have access to the failed terminal.

The most common way of coping with this problem is to install one or more extra terminal devices for each group of terminal users. In a bank branch, for example, six or seven tellers might each have a terminal, with an extra available for use in case one fails. Alternatively, it may be decided that a terminal failure could be managed by closing one of the teller windows until the device was fixed. Which of these approaches is chosen would depend on the level of business in the bank branch and the business impact of having a teller window out of operation for some time.

If only one or two terminal users form a group, the problem of backup becomes more difficult, although there are few real choices. If continued access to the computer-based system is essential, an extra terminal device is probably justified even though the cost may be relatively high for the system as a whole. If alternative methods can be devised to provide continued operation without a terminal (see the following discussion under Total-System Availability), the backup device can be eliminated.

Some systems provide more flexibility than others for the backup of terminal devices. This is especially true if terminals are used for different functions within the same physical area. For example, in a government office some people might use terminals to access a transaction-processing system while assisting clients of the agency. Within the same office, other people might use terminals for word processing, electronic mail, and similar office functions. It would be extremely useful if these terminals could be interchanged in case of problems, and especially if the terminals used for word processing could back up the client-use terminals—assuming that support of the agency's clients is the most important function. Whenever possible, this type of backup ought to be investigated, as it provides extra flexibility and can lower the cost of backup.

Total-System Availability

1. *Arrange for manual methods of operation in case of serious system failures, and in cases when terminal-device backup cannot be justified.*

In systems which serve main-line functions, the worst must always be anticipated. Terminal devices may fail, a major network failure might occur, a disaster might affect the main computer center. Still, business or government or health-care or educational functions must continue. As a final backup mechanism, manual methods for continued operation must be defined.

One way to handle terminal or network failures is to provide for voice-telephone input. Although this does not work very well if high volume is involved, it can be quite satisfactory for a small number of transactions. (If high-volume, continuous use of terminals is necessary, additional backup devices and facilities must be installed, to minimize the probability of complete failure.)

Another backup method is to provide hard-copy data from which remote users can work if they lose contact with the information processor(s) or if the processors are out of operation. In one bank, a printout of account status is obtained daily and kept until the next printout the following day. If the entire computer-based system fails, the tellers can work from the previous day's printout in cashing checks and authorizing account withdrawals. Transactions are batched and can be input to the system when it returns to operation. Although this method relies on somewhat out-of-date information, in practice not many people make multiple withdrawals in one day, and so the probability of error is relatively low. In addition, the personal judgment of the tellers and of the bank-branch management are relied on to determine if any transactions require extra attention.

The design-evaluation process (described in Chapters 14 and 15) must include a review of how the organization would cope with various types of information-system failures. This review will ensure that adequate planning has been done, so that essential functions can continue even when failures occur.

SECURITY AND PRIVACY PROTECTION

Security and privacy protection are important aspects of many computer-based systems. Although security protection and privacy protection have many different aspects and focus on different types of data, many of the same protection techniques are used in both. A specific system may not have any requirement for privacy protection if personal data are not stored or processed. However, almost every system has the need for security protection; the need is greater in some systems than in others, depending on the applications implemented and the possibilities for disclosure and/or fraud.

User Identification   The first step in protection is to identify the users of the system, because it is essential that only authorized users access protected functions and/or data. Personal identification can be achieved using one of the following methods.

1. *Arrange for each user to provide a personal identification, and/or possibly a group or project identification, when accessing the system.*

   The input of a user identification (ID) and possibly one or more passwords is the most common way to identify system users, but unfortunately it is far from foolproof. People in general do not protect their ID and password information; often they display this information on notes attached to a terminal, to a desk, on a bulletin board, and so on. Anyone who passes through the area can easily obtain a valid ID and password.

   In some systems it is also possible to obtain user ID and password information from computer-memory dumps. If the portion of the operating system which handles those checks is dumped, as it typically is after a total-system failure, the information is printed out on the dump. Even in installations with considerable sensitivity to security protection, system dumps may be treated carelessly, because the emphasis is on finding the problem and making the system operational again. There are methods to avoid disclosure from dumps; for example, the Honeywell Multics system stores ID and password information using a nonreversible encryption algorithm. Therefore, user IDs cannot be obtained by reading the dump or even by performing a decryption algorithm on data from the dump. It is also possible to use rotating and/or time-sensitive passwords, which are only valid during specific times or in a specific sequence. However, if the users write down this information and are careless about protecting the written copy, security is very poor.

2. *Consider the use of machine-readable badges or cards as a more positive personal-identification method.*

   Machine-readable badges or cards provide a more positive identification than key-entered IDs, although the positive identification is of the badge or card rather than of the person using it. However, it is usually more difficult to obtain another person's badge or card than it is an ID code, so that the protection level is somewhat better than with IDs. The use of a badge or card also cuts down on the number of keystrokes needed—one of the ground rules listed earlier for the good design of user-interface procedures.

Often the use of badges or cards for identification can be combined with physical-security measures to produce a high level of protection. For example, in many cases terminal devices are in locations where a stranger or an unauthorized person would be readily noticed and challenged. Bank branches are typical of this environment; even if an unauthorized person could obtain an identifying card (or personal ID and password), it would be impossible to obtain access to a terminal during working hours. Situations of this type probably do not require a high level of security protection in the computer-system hardware and software, because the physical situation provides the needed protection. However, care must be taken that off-hours use of a terminal is not allowed, as physical protection may be less effective during those times.

Similarly, some factory environments may be protected from outsider entrance, and the main reason for the use of personal badges is to log which person initiated each specific transaction (in this example, a by-product may be the record of jobs each worker performs, for payroll purposes).

3. *If a very high level of protection is required, more positive personal-identification methods are available, although at a high cost.*

The computer-industry literature describes many techniques for more positive personal identification. Examples include palm-print readers, fingerprint readers, and voice-print-identification devices. Each of these provides a very high level of positive personal identification, as the characteristics used for identification are almost impossible to forge. However, the technology in each case is relatively new, the demand is quite low, and the price is therefore high. This situation will change over time; the continually decreasing cost of microelectronics affects these devices as well as more typical computer components. However, the major factor will be increased demand, so that suppliers can achieve high-volume production. Even today, if the need for positive identification is urgent enough, it can probably be met through available technology of one of the types listed.

Access Control   The next aspect of security or privacy protection is to control access to sensitive data or functions. The key to control is, of course, the personal-identification process just discussed. Once the person requesting access has been identified to whatever degree of accuracy is necessary in the system, that identification can be used to determine what that person is allowed to access within the information system. The following two areas must be considered.

1. *Determine which functions must be protected and how the identification method(s) selected relate to controlling access to those functions.*

In some systems, the identification of a user allows that person to access anything within the computer system. In a single-purpose, single-application system this may be adequate. However, many large-scale systems are used for multiple purposes and shared by many applications. Allowing an identified user access to everything within the system is usually inappropriate. In today's systems, either of two possible methods can be used to ensure that each user accesses only allowed functions.

The most general method of control is to establish privilege lists for each user or for each related group of users with the same privileges. A privilege list, as the name implies, indicates which functions the user who is keyed to that list is allowed to access. For example, one user might be allowed to access only the time-sharing subsystem and within that subsystem to use only the BASIC compiler. Another user might be allowed access to several different query packages. A system operator or techniques programmer might have a privilege list including all the functions of the computer system.

A security administrator or perhaps a system administrator must establish and maintain the privilege lists. Of course it is important to update these each time a person changes jobs or leaves the organization (the oversight of not updating privilege lists, or user IDs and passwords, is the second most common security problem, after the display of personal ID and password information). Privilege lists entail a certain amount of overhead but provide a much higher level of protection than passwords do.

Another, more restrictive, method of control is to inflexibly link a user, or perhaps a terminal, to one type of function. For example, in a banking system a bank teller or teller terminal can be linked directly (in the system software) to the teller-support application programs. The teller, or any person using the terminal, has no legitimate need to access anything else within the computer system. Of course, that application must include all the necessary functions, such as training mode and "help" functions, which the user may need. In many transaction-processing systems this is the best way to ensure that users access only authorized functions—and through those functions, only authorized data.

2. *Define which data elements must be protected for security or privacy reasons and how this protection relates to the access control over system functions.*

Access to databases or files can be controlled using privilege lists directly, or indirectly through the allowed functions. If a specific function can access only certain data elements, no further access control is required. In other cases, additional control is needed, which can be provided by either a privilege list (which might be the same one used for functions or an additional one) or passwords or similar techniques. As in the earlier discussion, it must be noted that passwords—or any method which allows the user to input codes which control access—provide a very low level of protection.

Whenever possible, data access ought to be controlled via the functions accessed rather than independently. In many cases, logic built into the applications can provide the necessary protection. For example, if a salesperson in the furniture department of a retail store attempts to enter a transaction for a sale of men's clothing, either an error has occurred or an attempt is being made to misuse the system. The application logic can easily detect these situations, but of course they must be defined during system design.

One of the rules of security and privacy protection is to be careful not to disclose information when refusing access to unauthorized data. For example, if an unauthorized user attempts to access a record for which he or she does not have an access privilege, the system ought to return a "no such record" message, not one reading "unauthorized access." The former does not even disclose that such a record exists, while the latter may provide useful information to the person attempting to intrude.

Network Security   The next area of concern in security or privacy protection is how to manage the transmission of data over network facilities. The following two points must be considered.

1. *If sensitive information will be transmitted, the use of encryption devices must be considered.*

Financial data today are typically encrypted in applications such as the transfer of funds between banks. As more data which are sensitive from a business or governmental or a personal-privacy viewpoint are transmitted, the need for encryption will increase. Although this adds to the cost of data transmission, because encrypting and decrypting hardware is required at each end of each transmission link, it is generally worth the investment. Encryption at multiple levels may be needed if data are transmitted over a meshed network, in which misrouting might deliver messages to the wrong destination. Network encryption would not protect data from disclosure under those circumstances.

It is important to note that protecting against "eavesdropping" is not the only reason for transmission encryption; that is, encoding data not only protects against an outsider gaining access to those data but also makes it more difficult to tamper with message content or to add spurious messages to the transmission system. For example, if financial data were being transmitted in the clear (without encryption), a person tapping the link could modify data such as amounts or account numbers during transmission. Even more dangerous, a wiretapper could monitor transactions, determine the correct format and content, and then generate fictitious transactions such as a funds transfer to move money into a specified account.

Of course not only financial data require protection. Large manufacturing organizations, such as Boeing and General Motors, are beginning to exchange automated designs with their suppliers, using computer-to-computer data transmission. The design of a new product might be of great value to an eavesdropper, which might be either a firm competing with the major producer or a firm desiring to compete with existing suppliers to that producer. The financial-exposure potential of this type of intrusion may not be as immediately apparent as that of funds transfer, but under certain circumstances it could be as great.

Similarly, personal data transmitted over networks might be very sensitive and might also represent a source of potential financial gain through blackmail by unprincipled persons. Generally speaking, it is more difficult to put a financial value on the protection of personal data than on the protection of business or governmental data, so that the need must be established either because there are legal requirements for protection or because of public sensitivity to the problem—and therefore potential ill will generated if exposure occurs.

2. *Sensitive data transmitted over a PDN or VAN may present a potential problem, depending on the specific network.*

Data transmitted over a point-to-point private network can be encrypted very simply, using an encoding device when the bits go onto the link, with a compatible decoding device where the bits are received at the other end of the link. All bits transmitted are processed through the encryption and decryption devices.

If a PDN or VAN is used, however, this simple approach does not work. The network requires that header information, which includes the destination address and possibly other information used by the network-switching processors, be in the clear. If encryption is used, the encoding device must be intelligent enough to begin encoding only after the header data are transmitted, and the decoding device at the receiving end must similarly recognize and ignore the header. It is also possible that some PDNs and VANs will restrict their customers' ability to encrypt data, especially in the case of PDNs operated by PTTs. It may be necessary to use other transmission mechanisms if this presents a problem.

Physical Security   The final area of security and privacy concern in this abbreviated discussion is physical security. The entire question of protecting the computer environment is of course one which must be addressed, but it is adequately covered in the literature (see the Bibliography for references). The two aspects in which complex, distributed systems differ from more traditional data-processing systems are discussed here.

1. *Consider methods for controlling the physical security of dumps, file copies, and so on, whether these are in hard copy on paper, or in machine-readable form on diskette, disk, magnetic tape, or cassette. Any type of demountable media which can be detached from the computer system must be analyzed to determine the need for protection.*

It was pointed out earlier that security-protection methods may ignore items such as system-dump printouts which contain a great deal of sensitive information. In a central computer-room environment, those copies can usually be managed successfully, although the disposal of dumps is often overlooked in protection procedures. However, in a distributed system, where information processors are located in offices, factories, warehouses, and similar sites, care must be taken that memory or file dumps are not produced at those locations. If a dump of memory or file content is required, it ought to be up-line dumped to a central location for printout. In any case, there is typically no one at a satellite-processor location who can analyze the content of a dump, so its printout there would be pointless. However, some minis and micros have system software which will generate a printout dump if a failure occurs and a line printer is configured. In that case, the software must be modified to produce an up-line dump instead.

Similarly, the physical protection of any demountable devices at remote locations must be analyzed. Even the innocuous diskettes used on many word-processing systems may contain very sensitive information if company correspondence is created on the system. It is possible, of course, to become overly concerned about these aspects of security, but if valuable or sensitive data are processed on small systems of this type, methods must be found to provide an appropriate level of protection. In many cases the same kind of protection used for paper documents is adequate; for example, diskettes containing sensitive documents can be locked into filing cabinets. Often, however, when word-processing devices are introduced, the fact that diskette or cassette copies cannot be read directly by a human being makes people less concerned about security than with paper documents. Education is the key to providing an adequate level of protection in these cases.

2. *Study the overall physical environment of remote equipment for any needed protection.*

Remotely located equipment, unlike equipment in a central computer installation, may not be well protected physically. Of special concern is how the area is protected during off-hours, when the regular work force is not present. Many office buildings, for example, have after-hours cleaning workers who have full access to all areas. In addition, the cleaning force may not be subject to any type of security procedure, and there is no assurance that an intruder could not easily enter, masquerading as one of the cleaners. Of course this problem is not unique to the protection of remotely located computer equipment; many offices are poorly protected in terms of sensitive documents which may be readily accessible to cleaning workers or other building personnel. The potential for loss or damage may, however, become greater as computer equipment plays a larger role in the organization's functions.

Special security-protection measures may be necessary to ensure that satellite information processors, work stations, and their stored data are safe. It may, for example, be necessary to ensure that the areas in which those devices are located are cleaned during the day, when the work force is present, and that the area is secured at all other times. Special attention must be given to data stored at remote sites, as discussed earlier. It is hard to impose the same discipline used in computer centers in regard to handling tapes, disks, and so on, on personnel whose major job is not working with computers. However, security procedures may be necessary to protect the data resources as well as the equipment, as these may be of great value to the organization.

AUDITABILITY

Auditability is the final area which must be considered under integrity. Computer-based systems must be designed so that they can be audited, just as manual systems are. The main ground rule to be observed in this area is to involve the auditing staff—either the in-house auditors or the external auditors—in the system-design process. As in the case of the network administrators (see Chapter 12), if the auditors are involved in the design of the system, they will be in a position to suggest facilities which will later make their auditing function easier. Few system analysts have auditing background, so the assistance of experts in this field is essential. The following steps are important in designing for auditability.

1. *Design the system to be self-tracking, by logging all important events.*

A semipermanent log must be kept of all actions taken which affect the system and its operation. Lack of an adequate log is one of the most common failings in computer-based information systems, because designers and implementers tend to think of logs as adding unnecessary overhead. While it is true that keeping a log of system actions adds to the total system load, this is a necessary function which cannot be bypassed if the system is to be successfully audited. The items which must be recorded, either on a single system log or on separate, specialized logs, are as follows:

- Copies of transactions which have been processed as well as those which were rejected, with the reasons for rejection

- Database before and after images (required for auditing as well as for recovery and restoration purposes)

- Transaction responses, either full copies of all output messages or abbreviated records of where each message was sent

- Records of all actions taken by the system operators and administrators (for example, a system administrator's command to deactivate a certain type of transaction or to activate a new copy of an application program)

- Records of all automatic system actions, such as reconfiguration to drop a malfunctioning peripheral device

The log(s) must contain information about what occurs in the central information-processing site(s) as well as in the communications network and in any remote sites connected through that network. It is particularly important to be sure that the operating system automatically logs the actions of administrators and operators and if it does not do so, to implement additional logic to provide this function. This information is irreplaceable when determining what happened—both during problem diagnosis and in tracking down potential security breaches. It is, of course, true that a system which is known to be self-documenting may deter a potentially dishonest administrator, as the person knows that the actions taken will be recorded.

2. *Design for simplicity.*

This rule is a recurring theme of this book and is restated here to point out that simplicity aids in auditing as well as in system integrity (correctness and availability) and in flexibility (see next section). A system which is well organized and easy to understand can be audited much more easily than one which is a tangle of complex logic, poorly understood even by its designers and implementers.

DESIGNING FOR FLEXIBILITY

Many of the rules stated in the earlier parts of this chapter will, when followed, contribute to the flexibility of the system. There are several additional design guidelines which will help make it possible to change or expand the information system. These rules, although mainly oriented toward flexibility, will in turn contribute to the integrity of the system.

1. *Decouple information processing, database, and network design and implementation.*

This is a basic modularity principle—best stated by Hal B. Becker* and observed in the organization of this volume—which separates the discussions of how to design the information-processing functions (see Chapter 10), the database (see Chapter 11), and the network (see Chapter 12). Keeping each of these areas as independent as possible allows each to be changed without affecting any of the others. In fact, this is modularity on a systemwide basis and has the same effect as modularity at a lower level, such as within an application program. If this rule is not observed and functions associated with processing are intertwined with those associated with the network, for example, it may be impossible to change the organization of the applications without disturbing the communications network. The ability to change any part of a system without affecting any other part contributes greatly to flexibility.

2. *Decouple the design and management of the user interfaces, especially terminal-user interfaces, from the processing which uses the input data.*

Like many of the guidelines in this book, this is a well-known rule, but one which is too often ignored. It is quite easy to define input decomposition routines which handle the data supplied by the terminal user and present them to the application program. By interposing these routines between the input formats and the application, either the input

---

* *Functional Analysis of Information Networks*, Wiley-Interscience, 1973.

formats or the application logic can be changed without affecting the other. (Of course, if a major change to the user interfaces is made so that new data elements are added or existing data elements are dropped, both the interface and the application will be unavoidably affected.) This is particularly important when the users of the new system are unfamiliar with terminal usage; even though prototype interfaces are built and thoroughly reviewed with them, changes to the format or layout of the interfaces may be required as they start to use the system in production. The flexibility to make those changes can easily be gained by following this guideline.

3. *Standardize and document all interfaces between modules and programs.*

This is one of the ground rules of good structured programming practices and contributes greatly to the ability to change one module or program without affecting others with which it interacts.

4. *Limit the size of programs and modules.*

This is another basic rule of good structured programming which contributes to the ability to change—and to change quickly when necessary.

5. *Minimize the degree of tight integration among the parts of the information system; emphasize "loose coupling" instead.*

This is another of the themes which run through this book, but it cannot be repeated too often that integration reduces (or even removes) flexibility. Systems which are designed as a series of loosely related modules and programs can typically be modified, expanded, or contracted relatively easily. A system may be somewhat less efficient (in terms of computer resources and possibly even in human resources) when loosely coupled than when tightly integrated, but the trade-off in increased flexibility will almost always be well worthwhile.

6. **Don't** *define limits within modules, programs, or the system.*

One of the most common flaws in system design and programming is to include limits, such as the number of terminal users who can access the system simultaneously, the number of transaction types which will be used, the number of databases and/or files which one transaction will access, and so on. No matter how the limits are set and no matter how large the limit numbers may seem, they will inevitably be exceeded. As an example, one organization designed an order-entry system to handle all the orders for all the divisions of the company—even though only two divisions would initially use the computer-based system. In addition, the limits were set so that several additional divisions could be defined if a major reorganization of the company's structure occurred. There was confidence that the limits set would handle any possible situation. However, corporate management then proceeded to acquire another company, immediately doubling the number of divisions in the newly combined corporation. So much for careful planning in setting limits. It is much better to set none, although programs may suffer somewhat in efficiency if they are truly open-ended. Given the price/performance of today's computer hardware, this is a very attractive trade-off in return for increased flexibility to change and expand.

7. *Don't assume that the system will always be used in the ways which the system designers expect.*

This is a rule which every computer manufacturer has learned the hard way, through experience. A time-sharing executive system may be designed to allow users at terminal to create their own programs and execute them. Users may discover, however, that text-editing routines used to create programs can also be used for word processing, and soon secretaries are writing letters using time-sharing, professionals are writing theses and other papers using the same routines, and so on. Other users may discover that preprogrammed routines in time-sharing can be used for transaction processing, and soon salespeople are entering orders using time-sharing. None of these was an intended use when the executive system was designed and implemented, and these uses may uncover flaws which were not found when the use was only as envisioned by the implementers.

Examples of this type are far too numerous to relate. The key is to either design for great flexibility in usage methods or limit the use by very careful design—if it is important to limit users because of security, privacy, or efficiency concerns. (In many on-line systems these are very important aspects.) In addition, it is important to speculate about how the system could be used, given the proposed design, in ways not anticipated. A brainstorming session might be a good idea, with system designers and implementers not involved in the system being asked to speculate about possible misuses. Those can then be either guarded against or perhaps emphasized if it turns out that other possible uses would be advantageous. It is always a good idea to involve outsiders in reviews (this is the principle of structured walk-throughs), so that unbiased views are included.

Note that looking for unexpected methods of use and taking appropriate action to avoid or reinforce them will increase both flexibility and integrity. A system which is used in undefined ways is always a source of at least some uneasiness, particularly if sensitive functions or data are involved anywhere in the system. Removing this ambiguity will result in a better overall system design.

DESIGN EVALUATION

14

DEFINING
EVALUATION
CRITERIA

When the strategic-level system design is complete, it is necessary to evaluate that design, both to determine its ability to meet the requirements of the users and for technical feasibility. Design evaluation relates to the strategic-level system design as system test relates to system implementation. After the individual parts of a computer-based information system have been implemented and tested separately, they are integrated and tested together as a system. The purpose of that test is to determine if the system operates correctly and meets the defined requirements. Only after the system test has been performed successfully is the system ready for production use. Similarly, the strategic-level system design must be evaluated and determined to be adequate before the more detailed processes of design and implementation begin.

The system-design process, described in Chapters 9 through 13, results in a high-level design for the system interfaces, how the processing functions of the system will be structured, the proposed database structure, what type of network facilities and structure will be used to link together the system elements, and what methods will be used to ensure integrity and flexibility. The purpose of design evaluation is to determine if the design decisions were wisely made, and if not, to indicate how they must be changed. Iteration through some or all of the system-design steps may be necessary if the results of the design evaluation are unfavorable. It is, of course, preferable to iterate at this point in design and implementation, rather than to do so after a greater investment in the system has been made. The procedures outlined in this and the following chapter must not be bypassed, or an inadequate design may not be discovered until much later. This chapter describes how to define the criteria by which the success of the system design will be judged, while Chapter 15 describes how to evaluate the design against these criteria.

The design-evaluation procedures must determine if the proposed system will meet the users' needs and also if it is technically viable. Two types of evaluation criteria are therefore needed; one for user requirements and strategic objectives and another for technical aspects. The following section describes the criteria needed for the requirements evaluation, and the final section in this chapter describes the technical criteria. Both these sets of criteria will be used in the Chapter 15 process of design evaluation.

REQUIREMENTS-EVALUATION CRITERIA

The evaluation of the system design must determine if it will meet the needs expressed by the users, their management, the auditing staff, and any other group involved in defining system objectives. All the criteria necessary for this evaluation were defined earlier in system analysis. This section briefly summarizes the relevant criteria, with references to the chapters in which they are described in more depth.

• *Major application objectives*, in terms of what the new system is expected to achieve for the organization, are defined in the procedure described in Chapter 3. Examples of objectives which fall into this category include improving profitability, increasing efficiency, or meeting legal requirements. Design evaluation must clearly determine if those objectives will be met by the proposed system design. Often the designers lose sight of the strategic objectives in the process of defining more detailed user requirements, so it is important to evaluate how the system will measure up to its overall goals.

• *Functions to be performed*, described from the user's point of view rather than from a technical standpoint, are defined during the data-collection process in Chapter 5. This definition must include the types of transactions to be initiated by the users, the processing required to handle each type of transaction, and the resulting output to be prepared by the system. Functions which are not transaction-oriented, such as batch reports, ad hoc queries, and so on, must be included in this definition. Although it may appear unnecessary to determine that all required functions will be provided by the proposed design, in complex systems it is possible to overlook necessary functions— sometimes even until the system implementation is complete and production use begins. The evaluation process must therefore ensure that all the functions listed in the requirements statement will be provided.

- *Data to be stored by the system*, without regard to whether they will be stored in databases or files, are defined during data collection. Not only "typical" data but text or documents and also personal files to be stored must be included in this definition.

- The *response speed* needed for each type of transaction or report must be defined in the requirements statement. As the discussion in Chapter 5 points out, response is seldom defined as a single number but rather as a range of acceptable speeds, with a definition of how many transactions (or reports) must fall within that range and how many—if any— can be accepted outside those limits.

- *Turnaround speed* or *due-date requirements* must be defined for reports which are not appropriately defined in terms of response speed. Batch jobs typically are defined in terms of either turnaround speed or one or more scheduled due dates.

- *Availability* must be defined (during data collection) for each function within the system and possibly for each user group associated with each function. In addition, the requirements must include a statement of how availability will be measured, as this is not a simple process. Typically the first measurements of availability are made during system testing, with additional measurements during parallel operation (if any) and also at the time that cutover to production use occurs.

- *Integrity requirements* must be defined; these are described in Chapter 5 and discussed further in Chapter 13. Different levels of integrity for different functions and for different types of data and text may be appropriate.

- *Security- and/or privacy-protection requirements* must be specified, as described in Chapters 5 and 13, for different functions and for different types of data.

- *Auditability requirements* must be stated, in cooperation with the auditing staff. These requirements will vary, depending on the sensitivity of the functions carried out in the new system and also on the potential for financial misuse of the system. Chapters 5 and 13 discuss auditability requirements and guidelines.

- *Flexibility requirements* are stated during data collection and must include a definition of how this aspect of the system will be measured. The measurement of flexibility is not easy but can be achieved by specifying certain types of change which might be expected; for example, an increase in transaction rates, the addition of a new set of functions, or a reformatting of user interfaces. During the design-evaluation process it will then be possible to analyze how quickly and at what cost those changes could be made, given the proposed design.

- *Cost* is an important design criterion. When an application is selected as a candidate for implementation, this selection is often based on a projected ROI (see Chapter 2). That ROI is based on an estimated value of the application to the organization and on the expected cost to implement and operate the computer-based facilities. System-implementation costs include all expenses from the beginning of the analysis and design process through completion of system implementation, and usually also cover the costs of related activities, such as training for the system users. Operational costs are those incurred once production operation of the system begins.

  Operational costs are especially important. In a transaction-based system, these ought to be defined in terms of a cost per transaction as well as a total cost. The ability to measure and evaluate cost on a per-transaction basis allows the system to be monitored and analyzed in detail if it is not operating within the preset cost limits. Total costs are much more difficult to evaluate, because if these are too high it may be hard to decide which part of the system is causing the cost overrun. Some organizations may set more detailed cost guidelines; for example, network-facilities costs might be measured and evaluated separately from operators' salary costs, computer-equipment costs, and so on.

  The question of whether the proposed system design will provide the projected ROI (or better) is one of the key design-evaluation questions. At this stage in analysis and design, there is often a tendency to ignore the strategic-level goals and financial objectives, because by this time the proposed system has acquired a life of its own and a sort of legitimacy simply because it has been the subject of considerable effort for both the design staff and the users. This may lead to an attempt to downplay the need to meet the objectives originally stated, by pointing out all the intangible benefits to be obtained by implementing the system.

  Now is the time for a strictly business approach, not a sentimental decision based on the design staff's emotional investment in the new system. The cost objectives and limitations must be clearly stated in the evaluation criteria. If these are changed from the original statement (Chapter 2), it must be only with the approval of management at an appropriate level (based on the size of the sums involved).

- *Other criteria* may apply to an application within a specific environment. If so, these must also be defined and documented before design evaluation begins.

TECHNICAL-EVALUATION CRITERIA

The technical evaluation of the strategic-level design is equivalent to a structured walk-through at an early stage in the system life cycle. This evaluation must be carried out by a technically qualified team of people who are not directly involved in the analysis and design of the system being reviewed. Their purpose is to determine if the design will produce the desired results and if it is technically feasible.

The technical-evaluation process must use the same criteria as the requirements evaluation, because the ability of the design to meet those requirements is of paramount importance. However, the people who perform the technical evaluation must view the requirements and the design from a different perspective from the users. During design evaluation, the users are primarily interested in assuring themselves that their needs are fully understood by the system designers. Although the users are clearly interested in whether the proposed design can actually meet those needs, they are not usually qualified to judge its adequacy.

The technical-evaluation criteria will also consist of other factors which are not part of the user-requirements statement but which are germane to the review of technical feasibility. Some examples of items to be included in the technical criteria follow.

- *Conformance to the organization's objectives*, as defined in the discussion of long-range planning in Chapter 2, must be included in the design evaluation. It is especially important to determine if the proposed design follows the organization's philosophy for the use of local computing. The design process described in Chapters 10 through 12 may have resulted in decisions to distribute functions or data; those decisions may not be in accord with the overall system strategies of the organization.

- *Conformance to internal standards* must be evaluated at each review during system analysis, design, and implementation. During this first review of the strategic-level system design, few or none of the internal standards may be appropriate criteria. Many of those standards (see the discussion in Chapter 13) relate to programming languages, protocols, and interfaces, which become important later in the design process. Some standards, however, such as those which set guidelines for user interfaces, will probably be appropriate technical criteria for design evaluation.

- *Conformance to the rules of good design* is of course always a design-evaluation criterion. Chief among those rules are to design for modularity and simplicity, as restated in the first section of Chapter 13. One of the results of the technical evaluation must be an assessment of the degree of technical risk involved in continuing system design and implementation. Conformance to good-design rules is directly related to the decision which will be made regarding technical risk.

- *Other criteria* may be specified by the organization or may be defined for a specific system. For example, the organization may define the documentation needed at the end of strategic-level design; in that case the documentation rules must be included in the design-evaluation criteria.

When the technical criteria as well as the requirements criteria have been fully defined and documented, it is time to begin the design-evaluation procedure which is described in Chapter 15.

15

EVALUATION METHODS

There are many ways to evaluate a system design, ranging from relatively simple to very complex and costly. The appropriate method depends on the size and complexity of the application(s) to be implemented; the larger the investment in the new system, the more important a thorough design evaluation is. An adequate review of the system design is required, but the cost and time needed for evaluation must be kept within reasonable limits. This chapter describes two evaluation methods; one consists of design documentation and design reviews; the other also includes the preparation of system prototypes, models, or simulations. The latter allow in-depth investigation of how the projected system will operate and are appropriate when an unusually large or very technically complex system is planned.

DESIGN DOCUMENTATION

No matter how the system design is evaluated, complete documentation is essential, as this will form the base for the more detailed design and implementation steps which follow. The documentation which is required, as a minimum, is as follows.

• A *system description* must be the first item in the design documentation; this was first prepared, in narrative form, during the process of refining application objectives (see Chapter 6). If the description prepared at that time is still adequate, it can be used in the design documentation. Typically, however, changes occur during the strategic-level design process and must be reflected in the system description prior to design evaluation. It is also typical that as the system design progresses, the application is better understood by both designers and potential users, and the benefits to be provided can be more specifically identified. This is very useful during design evaluation because, as noted in Chapter 14, the designers may lose sight of the main objectives—which are to benefit the operations of the organization.

- *System flowcharts*, with accompanying narrative, must be provided. These are not the same as program flowcharts, but rather define the complete work flow, including both computer-aided functions and manual functions, so that it is possible to understand how the system as a whole will operate. The format used in Figure 5-3 is a good one for this purpose, and the initial set of flowcharts should have been prepared during the process of refining application objectives. Those charts must be updated to reflect the design decisions made during the processes described in Chapters 9 through 13, so that the flow of functions as envisioned in the design at this time is accurately described. As stated in Chapter 6, it is important not to use technical terms in these charts but, instead, to use terminology appropriate for user review. However, in a very complex system it may be necessary to prepare an additional set of charts, with more detailed technical annotations for the technical review. The remaining items which make up the system documentation (described in the following paragraphs) describe in more detail items shown on these charts and must be keyed back to the charts as appropriate.

- *System input formats and content* must be documented as a result of the interface-design process described in Chapter 9. At this point the design need not include all batch interfaces and possibly not all the interactive formats. However, the formats and content for all the main-line interactive transactions must be defined before design evaluation can proceed. Each type of input interaction can be documented as shown in the example in Figure 15-1, which is from an on-line order-entry system.

- *System output formats and content* must also be documented. As in the case of system input, batch-mode reports need not be completely defined at this time, although those which are important to the main system flow must be described. Interactive-mode output must be more fully defined, and output definitions for the main-line processing must be complete before design evaluation. The first definition of required output was obtained during the data-collection process described in Chapter 5 and documented as shown in the example in Figure 5-1. Changes may have been made to the requirements documented then, especially during the design of user interfaces and responses discussed in Chapter 9. If so, those changes must be reflected in the documenta-

System:	Order entry
Transaction:	Shipping-date change
Mode:	Menu and forms mode

User input	System response
Function key 2 (Order modification)	Display order modification menu (Form OE-4)
Select menu item 4 (Shipping-date change)	Display date change form (Form OE-6)
Enter order number and new date	Retrieve order record Display customer ID and name for verification
Enter "Y" or "N"—If "N" enter correct order number or terminate transaction; if "Y" processing continues	Check validity of new shipping date—if invalid, display error message
Correct shipping date or terminate	Change date in order record
	Display acknowledgement of change to user
	Print date-change notice to be mailed to customer and salesperson

Figure 15-1.   Interaction Documentation

tion used for design evaluation. The documentation of system output can be combined with input documentation if desired. This is usually appropriate for interactive modes of use, as input and output are so closely related.

• *Processing functions* to be provided may, in some cases, be documented in more detail than is shown on the system flowcharts. Those charts may be more easily understood if only a general description of functions is included, with more detailed descriptions as backup. The original documentation of processing functions ought to have been done during data collection, as shown in the example in Figure 5-3. That documentation must be modified if changes to the processing have been made during the other stages of design.

• *Data storage* to be provided by the system must be included in the documentation. During the data-collection phase, an initial set of documentation was created (refer to Figures 5-5 and 5-6). A more complete definition of database and file content and structure results from the database-design process described in Chapter 11. The documentation used in evaluation must include all the latter information, to provide as accurate a view of the design as possible. If design decisions have been made concerning how to organize the database(s) or file(s)—for example, to use a network structure, a relational structure, or an indexed structure—those decisions must also be documented prior to the technical evaluation.

- *Response speed projections* and how these have been calculated must be documented, because one of the design-evaluation criteria is the required response speed. A calculated response speed must be supplied for each main-line interactive transaction, accompanied by some indication of how the total system load is expected to affect the calculation. If a model or simulation of (parts of) the system has been constructed (see next section) and it includes an evaluation of response speed, complete documentation of the model or simulator, the input data used, and the results must be provided.

- Any proposed *integrity-protection methods* must also be described, covering all aspects of correctness, availability and survivability, security, privacy, and auditability. While it is desirable that these methods be documented so that they can be understood by the system users, it is most important that the descriptions be technically complete for use in the technical evaluation. Narrative methods are probably most appropriate for this documentation, although it may also be suitable to indicate on the system flowcharts the types of integrity features which will be provided at various points in the functional flow.

- *Flexibility provisions* are the final technical element to be included in the design documentation, and these—like the integrity methods—can be best decribed in narrative form. A matrix of the possible changes to be supported is also very useful in the format shown in Figure 15-2 (which is from an on-line-banking system).

Note that in several of the above cases references are made to documentation created during data collection and as part of the definition of application requirements. Those original documents must remain unmodified, although they can be used as the base for updated versions to be included in the design documentation. It is important to retain the documentation created during the entire analysis and design process, so that process can be tracked step by step. When this is done, any questions concerning when a change was made or why a format differs from that in the original documentation can be answered by reviewing the series of documents; it will be clear when the modification occurred. Trusting the

	Change transaction format	Change transaction content	Add personal ID check to transaction	Allow entry from automatic teller machine	Increase daily volume of transactions
New customer entry	Y	Y	N	N	Y
Checking account deposit	Y	N	Y	Y	Y
Check cashing	Y	N	Y	Y	Y
Savings deposit	Y	N	Y	Y	Y
Savings withdrawal	Y	N	Y	Y	Y
Customer record modification	Y	Y	N	N	Y

Figure 15-2. Matrix of Possible Changes

designers and/or users to remember exactly when and why changes were made is always a mistake. So much happens during analysis and design that no one can remember the details accurately; good documentation is essential.

COMPUTER-BASED EVALUATION AIDS

In addition to the design documentation used in the evaluation process, there are several computer-based aids which can be used. These include (1) prototypes of the system interfaces or of all or parts of the system functions, (2) a model of all or parts of the system, and (3) simulation of all or parts of the system. The purpose of using one or more of these methods is to ensure that the design is workable and that further investment in detailed design and implementation will not be wasted.

These computer-based evaluation aids are appropriate only for complex systems of broad scope, which represent a major investment by the organization. Any of these methods will involve additional cost, and the more complex methods (modeling or simulation) can be quite expensive. In fact, these methods are seldom used in application design because of their cost. However, if the planned application represents a significant investment—both in system hardware and implementation costs and in user training and the revision of operational procedures—it may be false economy not to spend the money necessary to ensure that the design is workable. In a very complex system, that assurance may be attainable only through modeling or simulation.

It is appropriate to point out that computer hardware, especially for high-capacity, large-scale systems, is designed and implemented using the techniques described here. The entire design is carefully documented, and important aspects are modeled and/or simulated. In addition, several engineering prototypes are typically built before the design is considered complete and released for production manufacturing. (The manufacturing organization then typically builds one or more pilot versions before cutting over to volume production.) This practice is universal among the manufacturers of computer hardware, because the investment in a new computer system is so large that it must be protected from design flaws. In relationship to the size of the organization, an on-line application may represent a comparable investment, yet these are almost never handled in the same way. Even the practice of building interface prototypes is far from universal. However, the same reasons for doing so exist as in the case of building new hardware systems—to minimize the risk and maximize the probability of success.

The main purpose of these evaluation aids is not to create a model, simulation, or prototype of the entire system; that is seldom necessary. The requirement is to focus on important functions and/or interfaces to ensure that their design is correct. An additional purpose is to determine the sensitivity of the proposed design to change. A prototype, model, or simulation can be changed relatively easily—at least in comparison to the complete system—and therefore the effects of various types of change can be evaluated before the system is implemented. This is especially important as main-line applications are implemented, because these change more often than batch-processing systems typically do.

Each of the three types of evaluation aids—prototypes, models, and simulations—is discussed in the remainder of this section.

PROTOTYPES

Prototypes have been discussed a number of times in this book, mainly in the context of building prototypes of terminal interfaces so that the users can evaluate them and suggest changes if necessary. If the most important user interfaces have not been implemented in prototype form earlier in the analysis and design process, this ought to be done before design evaluation. All the reasons for doing so were stated earlier and will not be repeated here. The importance of prototypes for users unfamiliar with terminals and computers cannot be overemphasized. The need for prototypes is an exception to the rule stated earlier—that computer-based evaluation aids are needed only if the investment in the new system is large. Even if the investment is moderate, the need to ensure that the interfaces will be acceptable to the users justifies the (usually small) cost of prototypes. It was also stated earlier that there is one exception to this rule: if all the users are already familiar with terminal-based systems, prototypes are generally unnecessary, as the users will be able to visualize the new interfaces from documentation and discussions with the designers.

In some cases it is also useful to build prototypes of some of the functions which the system will provide (this is in addition to the minimal functions required to support the interface prototypes). For example, a prototype would be helpful if the implementation will use a new type of database structure, such as a relational structure, with which the design staff has no practical experience. A prototype would also be useful if a new operating system, a new transaction executive, or other similar major piece of software is being used for the first time. A prototype of some of the planned application's functions is a good way to gain experience in how the software will react; that is, what its performance characteristics are, what the most suitable implementation techniques are, and so on.

One problem which may arise when implementing prototypes of parts of the new system is that there is often a terrible temptation to continue from the prototype directly into implementation of the entire system. Once coding begins, some implementers find it very difficult to stop and to return to the orderly design and implementation process which is so important in complex systems. It is never appropriate to implement the entire system to determine if the design is adequate; that is a very poor trade-off in terms of the effort expended. It may take a considerable amount of discipline on the part of the design and implementation team to avoid this temptation and to implement only prototypes of the essential parts of the system.

MODELS

Models may be appropriate to evaluate important parts of the system design. The creation of a model typically requires more time and effort than the implementation of a prototype and also requires skills which may not be available within the local staff. However, in a complex system a well-designed model can provide a much higher level of assurance that the design is workable than a prototype can.

A model is a mathematical representation of how the portion of the system being modeled will function. Often queuing theory is used in constructing a model, especially of a system such as a transaction-processing application. By making assumptions about the arrival rate of transactions, the probability that the database elements needed by each transaction will be immediately accessible, the amount of time it will take to process each transaction, and so on, it is possible to estimate how fast the response to each transaction will be—as well as the overall system load. Because each aspect of the system is represented by a number, a set of numbers, or a mathematical formula, it is possible to alter any one or a combination of these factors to determine what will happen if the assumptions are changed. For example, in a model of the type described, it is quick and easy to determine what effect an increase of 25 percent in total transactions will have and similarly easy to define what the result will be if twice as many instructions are required to process each transaction. Because of the need for flexibility, these estimates are very important in design evaluation. A model is simply a practical way to carry out estimates of system capacity and load much more rapidly than would be possible manually.

A model, like a prototype, is typically a rather broad-brush picture of the system or, more commonly, a part of the system. It reflects only certain major functions and relationships which are believed to be the most important in the system design. A model is very sensitive to the skill of the model designers because the results depend on the equations used as well as on the model's accuracy in reflecting the system design. It is therefore important that appropriate skills be available, or obtained from outside sources, before modeling is decided on as an evaluation method. A poorly designed model may give misleading results but may be viewed as being indicative of the system to be built, simply because it is a mathematical model. Some caution is clearly in order if modeling methods are to be used.

SIMULATION

Simulation is another way to mathematically describe all or part of a system, to evaluate how that system will perform. Simulation differs from modeling primarily in providing a more detailed view of system functions; a simulation is a more finely focused view of the system rather than a broad-brush picture as in modeling. However, the results of this increased detail is that simulations take longer to implement and also longer to execute than models do.

Simulation is also an area of expertise lacking in many IRM departments and is even more sensitive than modeling to the level of experience of the person who designs the simulation. Only in an extremely complex system whose overall feasibility may be in doubt, is it generally appropriate to use simulation. The detailed evaluation possible using simulation techniques may be necessary to ensure that the entire implementation effort will not be wasted. As in the case of modeling, a simulation is typically required for only the major functions and relationships within the system, especially since simulating an entire system would be extremely expensive and time-consuming.

Simulation can also indicate what will happen when changes are made to the system being simulated, and thereby predict how readily it will be able to adapt to new conditions. Simulation can make these predictions in more detail than modeling can, especially if the design is complex. If the person (or persons) designing the simulation is expert in the use of this technique, there is a high degree of probability that when the system is implemented it will perform as the simulation indicates.

DESIGN REVIEWS

Every design-evaluation process must include a minimum of two design reviews, one with the people who will use the new system and one with the technical-evaluation team. The focus of each of these reviews is different, but both are concerned with the ability of the design to meet the defined requirements and objectives. Although the review with users is perhaps more important, it is generally better to hold the technical review first, so that any technical weakness in the proposed design can be rectified before the presentation to the users and their management.

THE TECHNICAL REVIEW

The technical review must be carefully organized to meet the objective of a thorough analysis of the proposed design. This review can be described, in overview form, as consisting of a comparison between the design criteria (defined in Chapter 14) and the proposed design as reflected in design documentation, possibly augmented by prototypes, models, and/or simulations. The review is carried out by technically qualified individuals who are not directly involved in the system design being reviewed, to ensure objectivity.

The review team must be chosen to include the necessary specialities; in a complex system these are typically networking, database management, and distributed processing. Experience with office-support systems may also be important if the design being reviewed includes those functions, or may be expanded to include them in the future. Management must allow the necessary time to perform a thorough review; it is impractical to select a review team and ask them to continue full-time jobs while performing the review. They must be committed to the review process while it continues, which may be up to a week for a large system. Additional time before and after that week will also be necessary, to study the design documentation and record the findings of the review.

A review procedure used within Honeywell Information Systems for large, complex projects has proved quite successful; following is an outline of how that process operates. First, the review team is selected as described above. Each team member is provided with copies of all the relevant documentation, including the design-evaluation criteria (Chapter 14), the design documentation as described in this chapter, and documentation produced in conjunction with the implementation of prototypes, models, or simulations (if any). Each team member is expected to review this documentation in depth before the formal review begins.

The review itself varies in length, depending on the complexity and scope of the system being reviewed, but can last as long as a week for a large project. During the review, the system designers responsible for the project formally present the material included in the documentation as well as any other relevant information. Presentation may seem unnecessary if the review-team members have all studied the documentation. However, documentation is almost never unambiguous, and a presentation, accompanied by questions and answers, typically provides a better understanding than simply reading the documents does. Prior study of the documentation is, however, essential, as the team members would otherwise require much more detailed and lengthy presentations.

The review team must compare the design, as documented and presented, with the evaluation criteria defined and documented as described in Chapter 14. The technical background of the team members must ensure that they can adequately evaluate the design and determine if it will in fact meet those criteria. As in all structured design and programming reviews, there is, of course, no implication that the designers of the system are technically inadequate but rather that they become too close to the problem to see it in perspective. A team of uninvolved designers can typically be more objective than those directly responsible for the design and implementation.

Following the formal presentations, the review team may ask for any other information it requires and may also ask the presenters additional questions which may arise during discussions within the review team. The team in effect acts very much like a jury which deliberates on whether to find the design acceptable or unacceptable. The result of the review is a set of findings which are expressed as risks. For example, if the design documentation and the review findings indicate that interactive response speed will be 5 seconds instead of the 3 seconds stated in the evaluation criteria, a risk might be stated indicating that meeting the response-speed requirement will increase the cost of the system to an unacceptable level. Another risk might relate to the design's ability to change as required; the team's technical assessment might be that the design as proposed will be relatively inflexible.

Following the technical review, the system designers must formally respond to each risk, indicating what action will be taken to avoid each problem pointed out by the review team. There may be some problems which cannot be avoided, and management must then decide what action to take. For example, in practice it is seldom possible to meet all evaluation criteria, especially if they have been set without any concern for the practicalities of system implementation. The factor which most often causes problems is system cost. Many other criteria can be met if unlimited funds are expended on the system design and implementation. Response speed, for example, can be improved if high-speed links are used and/or if the capacity of the information processor and its disk files is increased. These increase the cost of the system and may make it exceed the cost goals established. Similarly, achieving high availability requires the use of additional hardware elements and possibly additional software and additional communications links, so that the cost of the system increases. Although readjustment of the design may minimize some of these problems, it is often impossible to absolutely meet all the evaluation criteria.

If the problem solution chosen is to ignore the cost criteria, the justification for the system, in terms of its ROI, may no longer exist. If the benefits to be obtained no longer outweigh the cost and the cost cannot be reduced through redesign, the ROI of the system will be negative—clearly an unpalatable situation. In this case, management ought to terminate all efforts on the system immediately, before further expenditures are made. It is a good ground rule that system costs increase during implementation, so that the final costs can be expected to exceed the estimates presented during design evaluation. There may, of course, be exceptions to this rule, but they are rare. If there is any doubt that costs can be kept within manageable limits, now is the time to kill the project and turn to another. Too often neither the system designers nor their management have the fortitude to make this decision, leading to continued design and implementation and often to an unsatisfactory, uneconomical system.

To look at another aspect of the cost problem, in some systems this is not a deciding factor. For example, some applications must be implemented or changed for legal reasons, and although the cost ought to be minimized, cost is not a legitimate reason for canceling the effort.

Only a design which is unworkable, either technically or because of its impact on the organization, would cause a reevaluation and a redesign effort. In other systems, the implementation costs, which are a one-time charge, may be excessive, but the real payback of the system is in its effect on operations over time. In that case, care must be taken to look at life-cycle costs when analyzing cost overruns. If the projected payback over time is still attractive, it may be appropriate to accept implementation costs which are higher than optimum in order to achieve that payback.

When evaluating the results of a technical review, each application and situation is different and must be judged appropriately. In many cases, the management of the using organization(s) must be consulted before a decision to terminate or reorient a design project is made. Those individuals are presumably familiar with the mainstream objectives of the organization and may be better able to judge whether meeting all the evaluation criteria is essential. They can also assist in making trade-offs among the various criteria, if all cannot be met. In many cases it is appropriate to enlist their aid when setting the evaluation criteria, so that priorities can be assigned. For example, if privacy protection is the most (or one of the most) important criteria, cost will probably be considered less important and added cost for the needed protection will be acceptable. If low cost is most important, other criteria such as availability or response speed may be sacrificed to keep costs down.

THE USER REVIEW

The user review follows successful completion of the technical review. Successful completion includes not only conducting the review, but responding to the risks generated during that review. The process of responding to risks may cause changes in the system-design documentation, and those changes must be reflected in the review with the system users.

Although the term "review" is used, it may be wise to hold several sessions with the users, so that the maximum number of people can be involved in the review process. This review ought to take a relatively short time, compared with the technical review. The designers present the system flow, functions, user interfaces, output to be provided, and benefits expected. A proposed schedule of implementation and cutover must also be provided. Questions and comments must be encouraged during the review, so that the users' views are obtained. Both individual users and their managers must be included in the review, either together or separately, as appropriate to the organization. In some cases, individual users will be reluctant to contribute comments or ask questions if their managers—especially high-level managers—are present. In other organizations the presence of the management may be supportive of user participation. These aspects are unique to the people and to the organization involved.

If prototypes of the user interfaces have been created, they ought to be demonstrated during the user review. If possible, users should be allowed to try out the prototypes as another dimension of the review. In many ways the user review is another facet of the process of selling the new system to its users, so that demonstrations, well-prepared visuals, and well-rehearsed presentations are appropriate.

Serious problems sometimes arise during the user review, leading to a reevaluation of the strategic-level design and iteration through some or all of the earlier analysis and design process. However, it is unlikely that real difficulties will be encountered at this point if the methods described in this book are followed. Problems during the user review are most often caused by lack of understanding of the users' requirements and/or by the failure to establish a good working relationship with the users.

Minor changes to the system design are often required as the result of users' comments or questions during the review. When those changes have been made to the design documentation, the strategic-level system design is ready for formal approval. The approval process is exactly the same as for the application objectives as described in Chapter 6 (see Figure 6-1). Approval by the manager(s) of the involved user group(s) is required first, followed by approval by the designated individuals from the IRM department. This approval is the next stage in the continuing contract between the users and the system designers and implementers, leading to the completion and use of the new computer-based information system.

SUMMARY

Design evaluation completes the strategic-level analysis and design process, which consists of the following steps:

1. Defining potential applications

2. Selecting a candidate application for study

3. Defining application objectives

4. Collecting data to describe the application's requirements and environment

5. Analyzing the collected data

6. Making high-level design decisions concerning:

- User interfaces

- System (information-processing) structure

- Database structure

- Network structure

- How to provide the needed level of integrity and flexibility

7. Evaluating the system design against the application requirements, and evaluating it for technical feasibility

When these steps have been successfully completed, detailed system design can begin, followed by implementation. The detailed system design can be accomplished using the same general flow described in Section 5 (System Design). Each of those processes is simply carried out in more depth.

The user interfaces, both input and output, are defined in detail for interactive and batch processes. The information-processing functions assigned to each processing location (assuming that a distributed-system structure was chosen) are broken into programs and modules, each of which is defined in detail. The exact structure, format, and content of each database segment and file are defined. Specific link types and speeds are selected for the network, the routes for the links are defined, and supporting software algorithms are designed (or obtained from an outside source). For each of these areas, the hardware and software features required for integrity and flexibility are defined in detail.

Throughout the design process it is essential that the same orderly methods described in this volume are used. Design reviews and project reviews must be scheduled at each major milestone in the project cycle. Milestones must be no more than 6 months apart, to avoid allowing the project to get out of control between reviews. For very complex projects, a 3-month cycle is even better. In addition, the design documentation must be kept up to date *and used* throughout design and implementation.

Using orderly methods during system analysis, design, and implementation is simply good common sense. Many of the methods defined for structured programming or design are based on the same themes presented in this book:

- Keep it simple.

- Break up complex problems into a series of smaller, simpler problems.

- Keep it flexible.

- Follow consistent methods.

- Prepare thorough, but understandable, usable documentation.

Unfortunately, a number of the structured methodologies hide these basic, commonsense guidelines under a cloud of technical jargon and/or bury them under (usually unnecessary) mathematical formulas. A good designer need not be a mathematician to cope with the majority of today's and tomorrow's computer-based information systems in business, government, or other organizations. A good designer must, however, cultivate orderly thinking and the ability to communicate effectively—not only with other designers and implementers but with the users of the system being designed. These abilities, the practice of good common sense, and observance of the principles presented in this book will produce a top-quality system analyst—with the additional potential to move into areas of the organization outside of Information Resource Management if a suitable opportunity arises.

CASE STUDY

16

ANALYSIS AND
DESIGN OF A
SAMPLE SYSTEM

This chapter describes the process of analysis and design in a sample system, which includes accepting orders for a company's products and taking the steps necessary for the items ordered to be shipped from the company's warehouses. This case study is actually drawn from two real systems so as to more fully illustrate the important aspects of system design. Some details have been changed, either to avoid specifically identifying a real system or to stress certain points. In addition, some detailed information has been omitted in order to limit this description to a reasonable size. The case study is presented in the same sequence as the preceding chapters of this book, which is the sequence of the analysis and design process.

SELECTING THE APPLICATION

Order entry was selected as a high-urgency application, because the company's business was expanding rapidly. As the number of orders grew, the manual system for handling them was becoming increasingly overloaded. Hiring additional order clerks was not a satisfactory solution to the problem, as the volume of paper being generated, tracked, and moved between the Order Department and the warehouses seemed impossible to control. A very sketchy appraisal of the potential ROI (see Chapter 2) indicated the following.

If the system remained based on manual methods, a 200 percent increase in the size of the order-processing staff would be needed to handle the expected 100 percent increase in the number of orders (because of the extreme inefficiency of the system when handling high volume). More people would also be required in each warehouse, in addition to those needed to pick and pack the stock, because orders were often sent to the wrong warehouse due to erroneous inventory records at the Order Department. Misrouted orders were sometimes sent back to the Order Department and sometimes to another warehouse which might have stock. In this uncontrolled environment, orders were frequently delayed and not infrequently lost. It was clear to executives in the Marketing Department that the company's competitive position would suffer and the order rate would fail to increase according to expectations unless these problems were solved.

However, even an optimistic estimate of the effort needed to implement a computer-based system showed that this could not be a short-term solution to the problem. A stock-inventory database had to be created first, before order entry could be implemented. Luckily, the company had been planning this move and had gone through the lengthy process of converting to the use of standard stock numbers at all locations. However, creating the database then verifying and loading the inventory position of all items could not be done overnight. In addition, of course, methods and procedures for computer-based order entry had to be worked out and the necessary application programs implemented.

After a preliminary study of the effort required, it was estimated that 18 months would be the shortest possible time for implementation, with the staff of 12 analysts and programmers available. More realistically, 24 months seemed to be required.

Given this situation, the management of the organizations involved—Marketing, Manufacturing, Distribution, Finance, and IRM—agreed on a plan of action which is shown graphically, in overview form, in Figure 16-1. This plan defined a phased design and implementation of, first, the order-entry application and then the programs to manage the paperwork in the warehouses. In parallel, a revision of the manual procedures for order handling would be undertaken, to keep the situation under better control until the computer-based system was ready. One of the IRM system analysts was assigned to participate in the definition of new manual methods, so that they could be made as compatible as possible with the methods being defined for the computer-based system.

Although the decision to revise the manual system was made for urgent business reasons, it was also an example of two good practices:

1. Provide a fallback position in case serious problems develop in the implementation of a computer-based system.

2. Avoid changes in manual procedures and the computerization of functions at the same time, whenever possible.

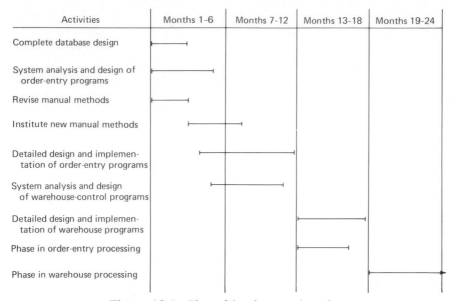

Activities	Months 1-6	Months 7-12	Months 13-18	Months 19-24

Figure 16-1   Phased implementation plan.

Unfortunately, provision is seldom made for fallback in case of delays or implementation failure. However, fallback ought to be considered in every complex system, especially when the technical risk is high. A phased implementation plan, as in this case study, helps to minimize the possibility of failure and makes it more likely that problems will be detected as they arise—not months later. Still, serious problems can occur, and ways to cope with them are always useful.

Because of the business urgency of the situation, only a minimal ROI analysis was made. By calculating the cost of the much-enlarged staff needed to (inadequately) handle the projected order rate and making a rough estimate of the lost business which was likely to result from order-processing delays and errors, it was easy to justify the expected cost of the computer-based system. In fact, however, the management decision to proceed was based less on the ROI analysis than on the belief that manual methods simply would not work. As long as the computer-based system was technically feasible and could be implemented by the available IRM staff (or even an augmented staff), management believed it to be necessary.

Often, of course, this type of decision is made only because of a good selling job by the IRM director or some other member of management. In the case-study example, it was probably the right decision, but a somewhat more extensive analysis of the amount of time and money to be invested, compared with the expected payback, would not have delayed the project unduly. Hasty decisions which involve large amounts of money are perhaps more common in relationship to computer-based systems than in any other activity of an organization.

At any rate, the company management decided to go ahead with the project, giving it top priority and postponing the implementation of any other major applications until its completion. Four system analysts were assigned to the project immediately, and four senior individuals from the functional areas were also assigned; two from Marketing (to which the Order Department reported) and one each from Manufacturing and Distribution. A middle-level manager from Marketing was given overall responsibility, as project manager, for both the revision of the manual methods and the installation of the computer-based system. Note that the Manufacturing Division was only indirectly involved in the first phases of design and implementation, but over the longer term it was management's intention to computerize more of the manufacturing-support functions and to tie those functions into the order-processing system.

DEFINING APPLICATION OBJECTIVES

The specification of application objectives (see Chapter 3), like the ROI analysis, was somewhat sketchy. The only formal definition of objectives was that contained in the internal announcement that the project team had been formed. That announcement stated that the team's objectives were to define new order-handling methods, both manual and computer-based, which would allow the company to handle its rapidly expanding order volume in an efficient and cost-effective way. A more formal and complete statement of objectives was prepared later, after data collection. This approach was undoubtedly correct, given the fact that the executives responsible for all the affected areas had agreed on the plan of action.

DATA COLLECTION

The process of data collection had to be coordinated with the definition of new manual methods, as those two activities were taking place concurrently. It was also the intent to design the new manual methods to be as compatible as practical with the eventual computer-based system. The first step, therefore, was to determine the flow of work and data in the current system. Through in-depth interviews with about a dozen salespeople and about a dozen members of the Order Department, the system designers defined the work flow as follows. (This description has been simplified by including only wholesale orders.)

1. Salespeople taking orders typically jotted them down in whatever form each person found convenient.

2. When in the office periodically, each salesperson phoned in the orders to the Order Department at headquarters. Order clerks transcribed the data received onto multipart order forms.

3. Some large customers phoned or mailed their orders directly to the Order Department, where they were transcribed by the order clerks.

4. All order forms then went to another group in the Order Department, who checked each order against customer-credit records. Any problems were referred to Finance for follow-up, while the remaining orders continued in the processing flow.

5. Still another group then received the order forms, and reviewed them against a file of ledger cards showing items stocked at each warehouse. If possible, the goal was to ship the entire order from the warehouse closest to the customer's location. If that warehouse normally stocked the requested item(s), the multipart order form was separated. One copy was retained in the Order Department, and the others were placed in an out box for courier or mail pickup. In cases of split orders, one or more additional order-form sets were created and the sets treated identically except for the warehouse to which each was sent. (The appropriate item(s) were included on each form.)

6. Orders arriving at a warehouse were date stamped but not otherwise registered. Each order was checked against a tub file of inventory-record ledger cards. For in-stock items, the inventory total was reduced and a picking ticket prepared. If there was no stock, either the order was back ordered, or for an urgent order, the clerk might telephone another warehouse to see if stock existed there. Sometimes, depending on the particular clerk's judgment, the order would be sent directly to the other warehouse or possibly back to the central Order Department for reprocessing.

7. The picking tickets were given to the stock pickers in the warehouse, and if the stock was there as expected, the ticket stub was annotated and returned for matching with the order form. The completed order form, with shipping information added, was then returned to the Order Department for billing. If no stock was found or inadequate stock was available, the picking ticket was returned and

matched with the order form. Further action was the same as in the original order receipt and again depended on the clerk's judgment.

This analysis of the work and data flow confirmed that the major flaws in the system were (1) the lack of standard procedures to be followed by all personnel; (2) inadequate tracking of the status of each order; and (3) inaccurate inventory records, even in the warehouses.

The system-design team decided that the first two problems required immediate attention and could be at least partially solved through improved manual methods. The computer-based system would complete the solution later. The team determined that the problem of tracking inventory could not practically be solved except by computerizing the inventory records, and they therefore decided not to attempt a manual solution.

After this initial analysis, the design team split into two groups, one to continue with the analysis and design of the computer-based applications and the other to revise the manual procedures. As was noted earlier, one system analyst was designated as liaison between the two groups, serving as a consultant to the functional (non-IRM) group redesigning the manual procedures, and feeding the results of that process into the ongoing system design.

To briefly summarize the results of the revised manual methods, the changes centered on the two problems identified earlier. First, standard procedures for handling orders in the warehouses were written, using three individuals from three different warehouses as consultants. Training sessions were held at each warehouse, and each person was required to complete the training successfully. Since these procedures improved control and standardized methods without affecting the system work flow, they could be implemented at each warehouse independently, as soon as the personnel there were trained.

Second, a formal order-tracking system was defined and instituted in the Order Department. Tickler files were set up, so that orders sent to each warehouse were not only identified but followed up by telephone if no shipment confirmation was received within a stated time. The new procedures at the warehouses ensured that orders which could not be filled were either back ordered, with a notice sent to the Order Department, or returned to the Order Department for rerouting to another warehouse.

Concurrently, a new, small-size order form was designed, which was more convenient for the salespeople to use in noting orders as they were taken. The data on the form was laid out to match the tentative design of the terminal-input format for the computer-based system (see User-Interface Design later in this chapter). The goal was to make the original orders more accurate and to allow them to be entered on terminals by the order clerks.

Meanwhile, the process of data collection (see Chapters 4 and 5) continued. The users of the system had been identified while the work flow was studied. The following user groups were potentially involved in the new system:

- Salespeople

- Order clerks

- Financial personnel (because of the need for customer-credit checks and customer billing)

- Stock clerks in each warehouse

- Stock pickers

- Management of the Order Department

- Management of the Finance Division

- Management of the Distribution Division

- Management of the Marketing Division

It had already been agreed that the system implementation would be phased, with order-entry functions implemented first and warehouse-control functions later. In addition, it was decided that because of the urgent need to implement operational-support functions, management-support functions would be deferred. It was hoped that if data elements likely to be needed were identified and included in the system database, necessary query and report facilities could be defined later. In fact, it was hoped that an ad hoc query and reporting package could be procured from an outside source, making in-house implementation unnecessary.

Because of these phasing decisions, the first part of data collection focused on the salespeople, order clerks, and financial personnel. It quickly became apparent that a basic choice must be made, which would affect a number of the application objectives. Order entry was currently an after-the-fact process, because the salespeople took orders then later phoned them in to the Order Department. In this environment, response speed was not crucial and system availability could be in the range of 95 percent or so—as long as outages were not unduly prolonged.

An attractive alternative was for the salesperson to enter the order from the customer's site, using the company's WATS (wide area telephone service) network. This would allow the salesperson to confirm the order to the customer on the spot (or indicate that items would be back ordered). However, response speed would then become a very important parameter, with 2 or 3 seconds being the maximum permissible time per item. In addition, availability above 95 percent (probably at least 98 percent) would be needed, with maximum outages of 5 minutes or even less. The system analysts involved were well aware that these differences in objectives would result in significant differences in the cost of the system. It was therefore decided that management must be consulted.

A review was held with the same management representatives from Marketing, Finance, Manufacturing, Distribution, and IRM who had made the original decision to proceed with the application. The alternatives, and their implications as far as they could be determined, were presented. After discussion, the management representatives again decided on a phased approach. All were agreed that direct order entry from the customer site was desirable, but, as had been tentatively decided earlier, this was too massive a procedural change to make immediately. However, they directed that this be considered a longer-term objective and that the system design include provision for evolution to this mode of operation. That decision allowed the data-collection process to continue.

OUTPUT REQUIREMENTS

Output requirements were next studied, and for the main-line interactive processes were identified as follows:

- Order acknowledgments, refusals, and picking tickets—all resulting from order-entry processing

- Customer-credit refusals, also resulting from order entry

- Responses to a variety of queries, which were not defined in detail at this stage

The data elements needed for each of these outputs were already well defined, because manually prepared forms for all these purposes (except inquiries) already existed. Although formats were likely to change in the new system, the data content would be the same. The existing forms were therefore included in the documentation prepared during data collection.

The response requirements, as initially defined, were for order-entry response ranging from 5 to 10 seconds, with no more than 3 percent of the total orders exceeding 10 seconds. The response speed was set based on the fact that order entry would continue to be after the fact rather than during the process of taking the order from the customer. It was agreed that if salespeople were to enter orders directly from the customer sites, response-speed requirements would change to a range of 2 to 3 seconds, with no more than 3 percent taking more than 5 seconds. The system design had to allow for the possibility of faster response requirements in the future.

Although the order clerks were not experienced in the use of terminals, the design team did not create prototypes of the user interfaces. However, they did arrange for a number of the clerks to visit a nearby company which, although in a different business, used an order-entry system very similar to the one being proposed. The clerks were able to talk to the order clerks in the other company and spend some time observing their use of the computer terminals. While perhaps not as good a solution as tailor-made prototypes, it served to familiarize the clerks with how a computer-based order system would operate and what their duties would be.

USAGE MODES

Usage modes were defined while output requirements were being considered. It was desirable that order-entry functions be as interactive as practical, and a variety of interactives queries was also required. The only batch-processing requirements defined initially were for weekly exception reports showing out-of-stock items with back orders which were not expected to be shipped for 3 months or more. Additional batch reports would probably be needed, but (as is often the case) it was decided that these could be defined during detailed system design.

PROCESSING REQUIREMENTS

Processing requirements were relatively easy to define, and the documentation of those requirements for the main-line flow of order entry is shown in Figure 16-2. The processing requirements for managing the warehouses were also investigated, although the intent was to implement these later. However, since orders had to be shipped from the warehouses, there was an extremely close tie-in with order entry. The processing flow for orders in the warehouse was therefore documented, as shown in Figure 16-3.

GEOGRAPHICAL LOCATIONS

Geographical locations were easily determined and documented. The most important locations were the company's headquarters, where the company executives and the Order Department were located, and the

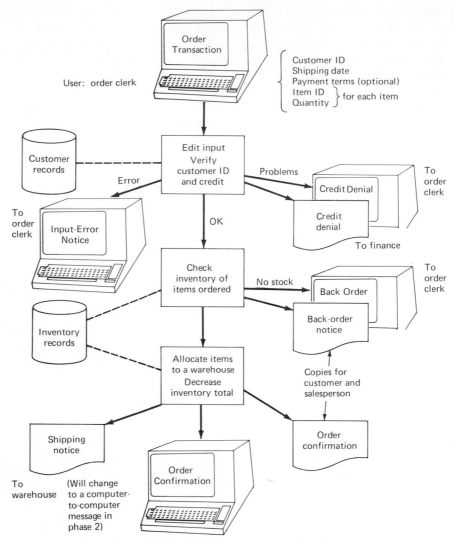

Figure 16-2 Order-entry-processing flow.

company warehouses, which were in twelve different cities around the continental United States (including the headquarters city). Although the salespeople would not be direct users of the systems, at least initially, their locations were also identified. These were the twenty-four regional sales offices from which the salespeople worked. These offices were also distributed around the continental United States.

The Order Department and sales offices were typical office environments, temperature controlled for comfortable working conditions. The warehouses were considered to fit into the "factory, normal" environment category (see Chapter 5), because they were subject to considerably greater ranges of temperature and humidity than the offices but did not include any problems of corrosive atmosphere, vibration, or other problems which would fit into the "factory, extreme" class.

Looking ahead to possible use of terminals directly by the salespeople,

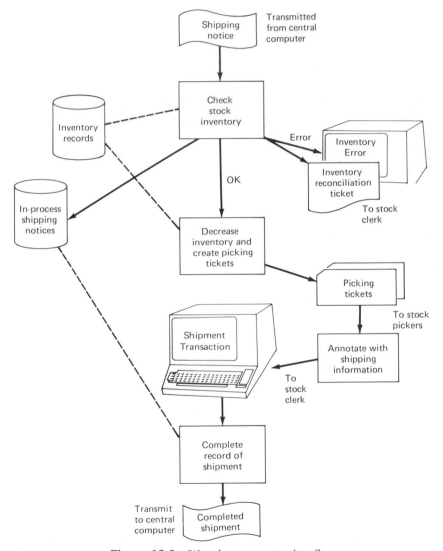

Figure 16-3   Warehouse-processing flow.

it was determined that if this were done, they would fall into the category of traveling users. The ideal long-term solution was for each salesperson to use a portable terminal to enter orders directly from the customer's site while taking the orders. It was decided that this would therefore be documented and carried in all system-design work as a long-term objective, so that if management decided to make the change to that mode of operation, plans would be in place for appropriate modifications to the computer-based system.

DATA-STORAGE REQUIREMENTS

Data-storage requirements were identified next. These included the basic information for customers of the company, as follows:

- Customer name and billing address

- Shipping address(es) if different from billing address

- Credit rating or limit

- Outstanding invoices and amounts not yet paid

- Outstanding orders not yet shipped

The other major set of data to be stored was the company's inventory, which included the following data elements:

- Item identifier

- Item description

- Item price

- Unit of shipment, if necessary

- Any special shipping instructions (e.g., fragile packing required)

- Inventory balance on hand at each warehouse (up to twelve warehouses, if the item was stocked at all)

- Item location(s) within each warehouse

- Back orders against this item

All the above information had to be maintained for the 43,000 items regularly stocked by the company. Because the company was expanding, as noted earlier, there was the probability that the number of items manufactured and therefore stocked would increase to at least 50,000, and possibly to 55,000. All planning therefore had to take these potential increases into account.

INTEGRITY, SECURITY, PRIVACY, AND AUDITABILITY REQUIREMENTS

Integrity, security, privacy, and auditability requirements were discussed with the users, their management, and the internal auditors. It was concluded that there were no special requirements for data correctness, although clearly the data to be processed and stored must be protected using the standard methods of validation, before and after journals, and so on. The requirement for system availability during the business day (starting at 9:00 A.M. eastern time and ending at 5:00 P.M. western time) was specified as allowing no more than one system outage a week, with a duration of no more than 15 minutes. This level of availability would make it unnecessary to have manual fallback procedures except for serious system outages—which could be expected to occur occasionally. Looking ahead to the possibility of entering orders directly from customer sites, it was agreed that the availability needed then would increase to no more than one outage a month, with a duration of no more than 5 minutes.

Security and privacy requirements were fairly normal. The customer information needed protection for both security and privacy reasons, while financial data and inventory balances would need security protection. Although there were no legal commitments to protect the customer information, the company felt an obligation to protect the customers' privacy. In addition, the fact that certain organizations were customers might be of interest to competitors of the company, so that a security aspect also existed.

Auditability was discussed extensively with the auditors. The main conclusion was that the system must be self-documenting and, to the degree possible, made secure from tampering with the self-documenting features. That is, the system must log each transaction and response, as well as each file-maintenance activity (such as changing prices, credit limits, etc.) and each action of the system administrators (such as changing the priorities of different transactions). This log would be maintained and printed out periodically on a printer in the Audit Department. In addition, the auditors would take part in each design and project review, so that they could become familiar with the system as it evolved and ask for any further auditing aids which were decided to be necessary.

EXPECTED CHANGE

Expected change was discussed with the users, their managers, the auditors, the IRM director, and the company's legal counsel. The results of those discussions were that the following types of change could be expected.

- Item prices changed frequently.

- Customers' credit ratings and limits changed occasionally.

- As noted earlier, the number of items manufactured and stocked was likely to increase.

- Also as noted, the number of orders being received was increasing rapidly and was expected to continue this growth for at least the following 2 years. A somewhat slower rate of growth could be expected thereafter.

- Management was considering building or acquiring a warehouse in at least one other city and possibly as many as three other cities, so that stock would be available closer to major customer locations.

- There was some possibility that management would decide to acquire another company in the same business, as a way to more quickly expand manufacturing capacity. However, it was not known which company might be the acquisition candidate, so that detailed planning for this possibility was not practical. However, as a minimum, this would cause the number of manufacturing locations to change, probably the number of warehouses would increase, and the number of items manufactured and sold might also change.

MANAGEMENT-CONTROL PHILOSOPHIES

Management-control philosophies were determined by discussions with the users' management in both the Order Department and the Distribution (warehouse) Division and also with marketing executives. The desire for centralized, tight control to avoid the anarchy which had grown up in the manual order-processing system was clear. In general, the management philosophy was in favor of distributed control with local profit and loss responsibility, but the future of the company was so closely tied to its ability to handle orders properly that this proved to be an exception to the general philosophy.

STATEMENT OF OBJECTIVES

The formal statement of objectives covered both the order-entry functions and the warehouse-control functions. Since the latter were not as fully defined as the former, they were described only generally. However, it was clear what functions would be provided (picking-ticket preparation and inventory management), because functional specialists from the warehouses were included in the design team. This was probably the most important factor in ensuring that the requirements for the later phase were fully considered. Without this level of involvement, a phased implementation might have resulted in the need to change the order-entry system to accommodate warehouse needs.

Only the system description for the case study is included here, for brevity, since the other requirements which make up the statement of objectives (see Chapter 6) were discussed earlier in this chapter. The system description was a narrative describing the system and its expected benefits, and read as follows:

> The new order-entry system for ABC Manufacturing Company will allow the company to handle customer orders in a more efficient, cost-effective way, by utilizing computers for repetitive work. Because the number of orders is growing at close to 40 percent per year, the new system is necessary if the goal of shipping each order within 1 week of receipt is to be met.
>
> The order-entry system will handle two major functions. It will control the processing of orders and keep track of each order from receipt, through shipment, to billing. (Later it will be expanded to handle customer payments as well, so that the entire order cycle can be mechanically tracked.) The system will also keep track of stock inventory at all twelve warehouses, both to support order entry and to assist warehouse management in controlling stock.
>
> The system will operate as follows.
>
> 1. Salespeople will record orders on the new order form (form number AD146). Each day the order forms will be sent by courier or Federal Express to the Order Department. If expediting is essential, the salesperson can telephone an order into the Order Department as at present.
>
> 2. In the Order Department, each order will be entered into the computer system, using terminal devices. The computer system will:
>
>   a. Check the customer's credit rating and refer any credit problems to Finance, through a printout in that department.
>
>   b. Check for available stock for each item ordered, using the same formula now used, so that all stock is shipped from the warehouse closest to the customer, if possible.
>
>   c. If no stock is available, a back order will be set up, with a notice returned to the Order Department. A printed notice will also be prepared, with copies to be mailed to the customer and the salesperson.
>
>   d. When stock is available, a notice to the appropriate warehouse or warehouses will be printed out and sent by the Order Department to the warehouse(s) by courier or Federal Express. Order

acknowledgements will also be printed for mailing to the customer and the salesperson.

3. Initially the warehouses will operate as at present, preparing picking tickets manually. In the second phase of the computer system, the picking tickets will be prepared automatically by the system.

4. In addition to the processing of orders, the computer system will provide a variety of inquiry capabilities, so that any customer, order, or inventory data managed by the system can be obtained by any person with a legitimate need for access.

5. The system will also provide methods for setting up new customer records and inventory records as well as for modifying or deleting customer, order, and inventory records. Only people whose jobs include the responsibility for these records will be able to use these functions.

The new computer system is expected to provide the following benefits:

• Make it possible to handle the growing order work load without increasing the Order Department and warehouse staffs proportionally. Over the next 3 years the workload is expected to grow by more than 100 percent; the goal is to hold the growth in the work force to no more than 25 percent.

• Increase the efficiency of order processing by maintaining control over in-process orders in the computer system and by keeping accurate inventory records in the system. It is the company's goal to ship every order within 1 week of its receipt.

• Improve the company's ability to react quickly to competition by making it easier to change prices and/or selling terms and conditions through use of the computer system.

The monetary savings which will be realized by containing workforce growth will easily pay for the computer system, so that it is fully justified on this basis alone. Increased efficiency and flexibility should also improve the company's competitive position, but the value of this improvement is difficult to quantify.

PATTERN ANALYSIS

The analysis of patterns (see Chapters 7 and 8) was the next step and was quite straightforward in this system. In fact, it presented almost a textbook example of how patterns can sometimes match preconceived notions of how a system ought to be designed. Ignoring for the moment management's data-access needs, there was less overlap among patterns than usual—contributing to a simple system design.

USER-GROUP PATTERNS

User-group patterns were distinct and exactly matched the groups of users identified earlier. The group patterns defined were:

- Salespeople—Their functional requirements were the ability to enter orders, to modify existing orders (including to cancel an order), and to query the status of existing orders.

- Order clerks—Their functional requirements included all of those listed for the salespeople plus the ability to create, modify, and delete customer records.

- Financial personnel—Their functional requirements included the ability to create, modify, query, and delete credit information in the customer records. They also needed to trigger customer billing on an exception basis, for specific shipments and/or customers.

- Stock clerks in the warehouses—These jobs would change radically when the inventory database was operational, as their main job was to manually maintain stock records. In the new system, many stock clerks would be reassigned to fill the need for more stock pickers. Those remaining as stock clerks would handle any problems with inventory balances. Their functional requirements would consist of querying and modifying inventory balances and location information (where stock was stored).

- Stock pickers—Their functional requirements were to determine where ordered items were stored and to resolve problems with the stock clerks if the items were not found where expected.

- Management—Their functional requirements were for access to a variety of data elements on demand. It had been decided earlier that these requirements would not be addressed in the first two phases of implementation. However, their data-access needs (see later discussion, under Management-Requirement Patterns) were studied to ensure that important data elements were not omitted during database design.

DATA-ACCESS
REQUIREMENT PATTERNS

Data-access requirement patterns for users other than management were pretty much as expected and closely matched the functional requirements of the user groups. Patterns identified were as follows:

- Order information was potentially used by the salespeople, order clerks, and financial personnel.

- Customer information was needed by the salespeople, order clerks, and financial personnel. Each salesperson required access only to his or her own customers' records.

- Inventory-quantity information was required directly by the stock clerks in the warehouse and indirectly (by order processing) by the salespeople, order clerks, and stock pickers. Only the warehouse personnel (stock clerks and stock pickers) needed access to "where-stored" information for each item.

GEOGRAPHICAL GROUPINGS

Geographical groupings were identified as follows:

- Groups of salespeople were located in twenty-four regional sales offices, spanning the continental United States.

- The order clerks and financial personnel were located in the same city and building but on different floors.

- Twelve warehouses were located in twelve different cities, spanning the United States. All but one of the warehouses were located in the same city as one of the regional sales offices but, in each case, in a totally different part of the city. One warehouse (and one regional sales office) was located in the same city as the Order Department.

- All the executive-level (potential) users of the system were located in the same city as the order clerks and financial personnel but with offices in two different buildings.

INTEGRITY-PROTECTION PATTERNS

Integrity-protection patterns were not found in this application. It was determined that some fields, such as inventory-item description, could tolerate a lower level of accuracy than fields such as inventory balances; however, there were too few of these to practically segregate from the remaining data.

SECURITY AND PRIVACY PATTERNS

Security and privacy patterns were identified as follows:

- Security mechanisms were needed to control access to and modification of customer credit and billing information, item prices, and item balances, because improper modification of any of these data might allow fraud.

- Access to customer information required control for both security and privacy, as defined during data collection.

MANAGEMENT-REQUIREMENT PATTERNS

Management-requirement patterns related to both data access and control and were identified as follows:

- Order Department management needed access to all data related to orders, both current and historical. They wanted to be able to access records of individual orders for a minimum of 3 months after completion of shipping and billing. They needed access to statistical summaries of orders for a minimum of 3 years and would prefer a 5-year period (to match the company's 5-year planning cycle).

- Finance Department management wanted to access all current customer and order data and also to maintain their customer billing and payment histories for a 2-year period. (There was no immediate intent to include customer payments in the computer-based system, but it was obvious that this extension would be desirable.)

- Distribution Division management were involved at two levels, the warehouse and the division. Each warehouse manager needed access to all data about current inventory and order-processing status there, as well as inventory-usage-trend data (for at least 2 years) and order-trend data. Divisional management needed only trend data, but for each warehouse plus consolidations for any specified set of warehouses and for a 5-year period.

- Marketing Division management needed access to any specific customer or order data on request, plus customer-, item-, and order-trend data for 5 years.

Note that the data-access patterns for all divisional-level management overlapped and that all were heavily dependent on historical data to determine trends. Lower-level management requirements were more oriented toward current data, with some shorter-term trend data. This was consistent with the range of duties at each level.

- Management-control requirements were very clearly stated; uniform methods and procedures must be ensured throughout the organization. Management's experience with the manual order-processing system had convinced them that locally defined methods were very disruptive and that efficiency could result only from standard methods. This translated into two rules for the computer-based system: (1) all programming would continue to be done by the central IRM staff—although approved query and reporting software would be made available to authorized individuals; (2) the procedures to be carried out would be computerized to the maximum degree feasible, to ensure consistency.

USER-INTERFACE DESIGN

While pattern analysis was going on, other system analysts were designing the user interfaces for the order-entry phase of the system. This design was integrated with the definition of revised manual methods, including the design of the smaller order form mentioned earlier.

As a result of this integrated approach, the order form and order-entry-screen format were almost identical, as shown in Figures 16-4 and 16-5. This made it very easy for the order clerks to transcribe data from the forms onto the terminals. In addition, it made it practical to consider direct terminal entry of orders by the salespeople in the future, because anyone familiar with the order form would find it reasonably easy to use

ABC Manufacturing Company
Order

Customer Name ____________________ | Shipment Date ________

Customer ID ____________
(for a new customer, submit Form AN44 to obtain a customer ID)

Payment Terms
(enter only if nonstandard)

Items Ordered

Item ID	Quantity	Item Description (information only)

Continue on Form AD146A for more items

Salesperson Signature

Form AD146

Figure 16-4 Order form.

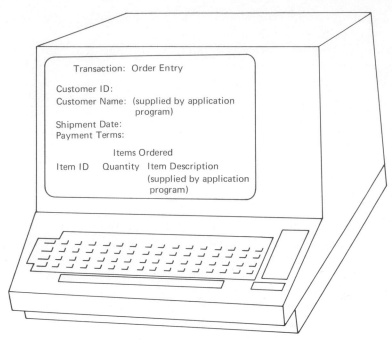

Figure 16-5 Order-entry screen.

a terminal instead. Because the order clerks were not, in general, trained typists, the same interfaces which they found satisfactory could be used by the salespeople.

The approach chosen would also allow for an intermediate step, in which salespeople would telephone orders to the order center, as described earlier, for terminal entry by the order clerks. This would provide immediate order entry and response but without changing the roles of the salespeople and order clerks. The management and system analysts in this case followed two rules of good design:

- Minimize the change (and potential disruption) at each phase of system implementation.

- Preserve the utmost possible flexibility for future evolution, both in computer-based functions and in manual methods.

These rules were particularly important because the company was expanding rapidly in a highly competitive industry. Some future requirements, such as the need to handle higher volume, were known, but others could arise unexpectedly because of changes in the market or because of moves by competition. The emphasis which management placed on flexibility was therefore appropriate.

One other user interface was defined at this stage of system design; a general-purpose inquiry. Originally the intent was to design specialized inquiries for the user needs defined, but this proved to be time-consuming and unnecessary. Instead, a menu-driven system with a fast-track option (see Chapter 9, Figure 9-3) was chosen. The user could press the function key associated with queries to obtain the first screen, which offered a fast-track option. An experienced user could then enter the necessary query data, while other users progressed to the next menu. That menu listed the possible queries—basic customer information, customer financial data, item-inventory information, item pricing, and order status.

By selecting one of the options on the menu, the user moved to the final menu, asking for selection criteria for the specific record(s) desired. Because there were requirements for security and privacy, controls were built into each menu. For example, if the choice of the customer financial query was made, the query package would check to determine if the selection originated at a terminal physically located in the Finance Division and also that the system calender and clock showed it to be during the normal business workday. If either of these checks failed, the query was refused and the situation was reported on a system administrator's VDT. An authorized administrator could override these checks temporarily if, for example, someone from Finance was working overtime or using a terminal at another location. In this particular application, query refusals for security or privacy reasons were not concerned with revealing information by the refusal itself (see the discussion in Chapter 13), because all the information which could be asked for was obviously available in the system.

Similar checks were included in the final menu. Because the system analysts were aware of the limitations of user ID and password controls, they relied whenever possible on physical security. All attempts to bypass security and privacy controls were displayed and also permanently logged, so that any patterns of attempts could be analyzed.

The remaining user interfaces, including batch reports, management queries, and the interfaces related to the warehouse-support phase of the system, were defined later. For brevity, they are omitted here.

CHOOSING THE SYSTEM STRUCTURE

When pattern analysis was complete, the system structure had to be chosen. As in many applications, not all of the decision criteria (see Chapter 10) pointed to the same choice. This was especially true when the later phases of the planned (and potential) evolution were considered. The criteria studied included the following.

• *Capacity studies* indicated that a single computer system would not be able to handle the total system load, especially as other functions (beyond those initially defined) would clearly be added. Also, rapid growth meant that a considerable capacity reserve would be needed. This factor indicated that a distributed system must be considered.

• *Flexibility* was a high-priority requirement, as was high *availability*. Both these factors pointed to distribution.

• *Interaction* was stressed, to improve the efficiency of all system users. This also indicated a distributed-system choice.

• *Clustered functions* were prominent in this system, leading to a distributed structure. *Customization*, however, was not a stated requirement, and in fact management was strongly in favor of standardization.

• *Security and privacy protection requirements* existed but were not considered stringent enough to force centralization.

• *Geographically*, the total set of users was quite dispersed, especially if the salespeople were considered as candidate terminal users. However, the order clerks—the main users initially of the order-entry application—were centrally located. This tended to favor centralization of their support functions but distribution of functions such as warehouse support.

• *Loosely coupled or undefined functions* were not a major factor, at least initially. The first functions to be implemented were well understood. It was expected, however, that when manufacturing-control functions were implemented later, these would be loosely coupled with the order-entry and inventory-management functions.

• *Management style* favored centralization, or at least centralized control. Since other factors indicated that full centralization was not a viable alternative, control would have to be emphasized during system design.

• *Technical risk* would be greater in a distributed system, although two of the system analysts had worked with a distributed system at another company. They had also had some experience with a very simple distributed database. However, the risk was clearly a matter of concern and was to be carefully managed.

• *Cost* was not analyzed by comparing the cost of a distributed versus a centralized approach but, instead, was analyzed in terms of ROI. As stated earlier, a very rough estimate of cost, assuming a distributed-system approach, could easily be justified. This was far more important than whether a distributed system would be more expensive than a centralized one. In any case, the latter choice was not a valid one for the system as a whole.

The tentative decision made at this point, based on these criteria, was to centralize the major functions associated with managing customer accounts and accepting orders but to distribute as many of the warehouse-management functions as practical. Further study of how to handle the database and how to structure the network (see the following discussions) would either reinforce this decision or lead to its revision.

DATABASE DESIGN

The next step was the high-level design of the database. Realistically, the database had been given a lot of attention even before the decision to proceed with a computer-based order-entry system. It had been apparent for some time that the inventory would have to be moved onto a database, so various alternatives had been studied. As a result, this part of the strategic-level design was more advanced than is typically the case.

The key factor in database design was the patterns of data access identified earlier. Ignoring for the moment management's needs (as had been agreed), the total set of data elements could be logically grouped as follows:

- Basic customer data and credit ratings

- Customer billing information

- Order information

- Item inventory for each warehouse

- Item-location information for each warehouse

Because multiple user groups required access to customer data, there seemed to be no logical way to partition those records. Each salesperson required access to only his or her customers' data, but splitting up the database in that way would serve no purpose, especially as the supporting functions were best centralized. Customer billing information could best be centralized for similar reasons. The customer-related data would therefore form a centralized database or database segment.

Order information, which required links to the associated customer data, was also best suited to centralization. This would be either a part of the same database used for customer information or possibly a separate database with some form of linkage to the customer records.

Inventory records presented a more difficult set of design choices. The data associated with item locations within a warehouse were clearly needed only in that warehouse. Item prices were needed only centrally, not in the warehouses. Both item identifiers and inventory totals were, however, needed at both the warehouse and at the central site (for order processing). The question was whether to carry this information at only one location—and if so, at which one—or to duplicate it at both.

A relatively quick trade-off study showed that item identifiers were essential at both the central location and the warehouses. The item description, however, was not really needed in the warehouses, as long as it was available on request from the central site. After much discussion, the design team decided to replicate inventory balances at the central and warehouse sites. At each warehouse, the important information was exactly how many of each item were stored in each warehouse location (a given item might be stocked in more than one area of the warehouse). At the central order-processing location, in contrast, the exact location of the items was unimportant, but the amount of stock in each warehouse formed the basis for deciding how to allocate orders for shipment. The design team envisioned that the inventory records would be handled as follows.

Each warehouse would maintain its own database showing the number of items of each type stored in each warehouse location. A total of the stock for each item in each warehouse would be carried in the central database, so that the total stock in all warehouses could be easily determined. The total for any item in any warehouse in the central database should, at any time, equal the combined total for that item (perhaps in multiple locations) in the warehouse database.

During order processing, the totals in the central database would be used to determine where to send the order for shipment. If, for example, item 1234 was ordered by a customer close to warehouse 4 and an adequate stock of item 1234 was on hand there, the order would be sent to warehouse 4. The central database inventory record for that item would be decreased in the warehouse 4 total, and a message would be sent to that warehouse indicating the order number, stock required, ship-to address, and any other necessary information. The message was really a partial picking ticket, which was completed upon receipt at the warehouse by adding the necessary information about where the stock was located. As each such message was processed, the stock quantity in the warehouse database would be reduced. To avoid getting out of synchronization with the master database, the messages to each warehouse would be sequence numbered, so that they would be processed in the same sequence as at the central location.

Although in theory this process would work correctly, it was recognized that the balances might not remain in synchronization. As a result, any difficulty in processing the picking tickets or in picking the stock would cause the warehouse stock clerks to be notified. It would be their responsibility to reconcile the inventory balances in the warehouse and central database, adjusting whichever was incorrect to match the true stock total. In addition, provision was made for a periodic synchronization of the balances, as rolling physical-inventory counts were made in the warehouses. Each such count would update both the warehouse database and the appropriate entry in the central database, although for auditing purposes the status of the inventory total before correction would be included on the system log.

This database design violates some of the principles of good distributed-database design, because totals are maintained at two locations and updated semi-independently in parallel. However, the design team considered other choices which would not have this drawback and found that each had other difficulties. For example, the warehouse managers felt that if all balances were kept only at the central location, their responsibility for control of the stock balances would be undermined. (Note that this was a psychological problem more than a real one, but was extremely important nonetheless.) If the balances were kept only at the warehouses, there would be a great deal of traffic back and forth while the order-entry system tried to discover where to send orders (as was the case in the manual system). The design team therefore decided that the solution chosen was the best, given the circumstances. They agreed to design extra checks to ensure that the problems of data synchronization would be manageable.

NETWORK DESIGN

High-level design of the network (see Chapter 12) came next. It would be necessary, as soon as the warehouses were linked into the order-entry system, to connect all twelve of the warehouses to the central order-processing system. Later it would probably be necessary to extend the network to include all twenty-four regional sales offices; however, it was decided to postpone consideration of that network extension.

Eleven of the twelve warehouses were located in cities remote from headquarters, while the twelfth was in the same city. Because the headquarters city was on the west coast, communications to and from the other eleven cities would probably be expensive. An initial decision was made that dial-up connections presented too many potential problems (errors and the inability to establish a connection at some times); therefore dedicated links would be used even if traffic volume did not justify this choice.

The first trial network design, shown in Figure 16-6, connected each warehouse directly to headquarters. However, the cost of this network would be extremely high, and the volume of traffic to and from the warehouses did not make a dedicated link necessary in each case. The advantage of this design was its simplicity, plus the extra capacity which would make network expansion unnecessary even if order volume grew far more than expected.

Two possibilities for reducing network cost were then considered: one was to install a meshed, packet-switched network. Although this would lower network operating costs, as well as provide alternate-routing capabilities, the staff had no experience in implementing or operating this type of network. They therefore decided that the risk of failure would be high, even if a turnkey implementation were contracted. The importance of the system to the company left little room for high-risk approaches, so they turned to the other alternative.

This was to install a multipoint network, as shown in Figure 16-7, to connect all the warehouses to headquarters. Although multipoint links are not often used to connect multiple computers together, there is no reason why they cannot be. They provide the same reduced cost in this

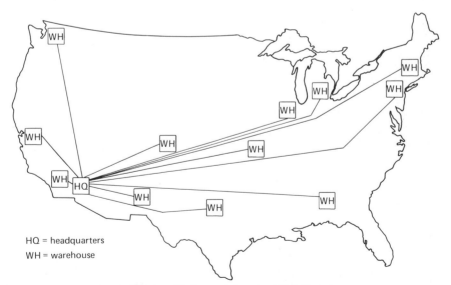

HQ = headquarters
WH = warehouse

Figure 16-6   Trial network design.

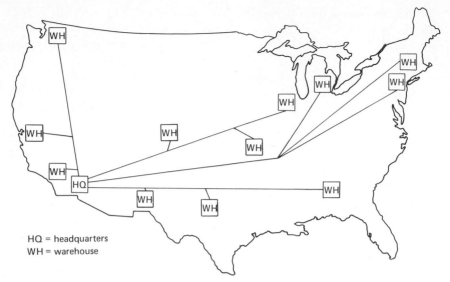

Figure 16-7 Final network design.

case as when used for the more usual purpose of connecting multiple terminals to a computer.

This suggestion reduced the estimated cost of the network considerably, even allowing for the use of high-speed modems to obtain the fastest possible transmission speed on the links. Because several programmers on the staff had extensive experience with multipoint links (used with terminals), this approach reduced the risk considerably—even though some changes would be needed in the vendor's software to manage the multipoint connections. (The vendor agreed to make these changes on a contract basis.)

The network design chosen was valid for that time. Today, however, both the GTE Telenet and Tymnet value-added networks serve all the locations in this network as well as the locations of the regional sales offices. If that had been the case when the network was designed, use of one of these VANs might have been even more cost-effective than the approach chosen. It is possible that the network will, over time, evolve into a hybrid combination of dedicated links and VAN connections.

INTEGRITY AND FLEXIBILITY

Considerations of integrity and flexibility (see Chapter 13) were addressed as an integral part of the strategic-level design, rather than as a separate step. The results are summarized here.

CORRECTNESS

Correctness was emphasized throughout, and perhaps most prominently, by assigning as much of the order processing as possible to the computerized applications. Minimizing the amount of data entry, the manual processing required, and manual decision making were key elements in ensuring correctness. Database integrity was ensured by rigid controls over access and modification as well as by the use of DBMS software which included concurrency controls, journal creation, and automatic recovery.

Additional controls over the correctness of the distributed database were required, but those were implemented as application routines and database-administrator aids, to avoid the need to change the vendor's DBMS software. This was another decision made to minimize risk, as changing complex software may cause problems, and the vendor may be unwilling to fully support locally modified software. These additional capabilities also provided for database recovery and synchronization (part of availability).

Transmission correctness requirements led to the use of the HDLC protocol. In addition, application-level controls were designed to ensure the correctness of transmitted data where this was especially important.

AVAILABILITY

Availability received a lot of attention. The existing central computer had full automatic recovery capabilities as well as the ability to gracefully degrade operations if a component failed, but the minicomputers considered most closely suited to the warehouse requirements could not operate unattended. Their manufacturer was requested to quote a price for modifying the software (and hardware, if necessary) to provide that capability. The resulting quote was reasonable, leading to the eventual selection of that vendor's equipment.

Network fallback was studied in depth. The choice of a multipoint network complicated availability somewhat. However, it was determined that dial-up capabilities could be provided at each warehouse mini by installing an extra modem. This would allow the warehouse system to reestablish connection with the central site if either its link or its main modem failed. Special software in the central system was necessary to handle these switchover situations.

Various combinations of problems were studied in an analysis of total-system availability. Because the computer-based system so closely paralleled the manual system, fallback to a completely manual base would be feasible. As long as either the central system or its local system was operating, each warehouse could continue processing orders. Both inventory balances and where-stored information for each warehouse were periodically to be dumped onto magnetic tape. (The where-stored data was to be transmitted on a cyclic basis to the central location for dumping at night.) Those tapes would serve as emergency backup, so that the data could be printed out and used to locate and pick stock if total failure occurred at one or more locations. It was believed (and later proved in operation) that serious failures would slow down order processing but would not stop it.

SECURITY AND PRIVACY

Security and privacy, although important, were not major concerns. Because the system was interactive and transaction-based and used terminals in known locations, the necessary controls could be provided in the application programs. Each terminal user could be positively limited to the functions and data elements legitimately needed. The controls used were verified by the audit staff, who also conducted periodic reviews that included attempts to penetrate the system. Those reviews were intended to ensure that a programmer or system administrator did not dismantle the controls during system operation.

Network security was not considered to be a requirement, so no special action was taken.

PHYSICAL SECURITY

Physical security in the warehouses was a cause of some concern, largely because the computer equipment used there would be vulnerable to sabotage. It was considered possible, though unlikely, that dissatisfied workers might vandalize the computer systems. This problem was avoided by installing each mini in the warehouse manager's office area and providing locks to secure the area properly at night.

AUDITABILITY

Auditability provisions have been described already. The auditors worked very closely with the design team and continued to monitor the system and its operation over time. This approach ensured both that the system was operating correctly and that it was not being subverted in any way.

FLEXIBILITY

Flexibility had been defined as one of management's major goals for the system, so this was emphasized throughout analysis and design. Structured design and programming methods would be used. Limits would not be set in any of the application programs. Every design review, beginning with the design-evaluation process described next, included flexibility as one of the review criteria.

DESIGN EVALUATION

The design-evaluation criteria (see Chapter 14) consisted of the statement of objectives accompanied by the requirements defined as a result of data collection. There were also a few criteria defined for technical evaluation, but this company (like many others) had no long-term guidelines or detailed standards. As a result, the technical-review team used their best judgment rather than well-defined standards in determining if the project was in conformance with company objectives and practices.

Technical evaluation of the design (see Chapter 15) was performed by three system analysts not involved with the application. No formal review was held; instead, the review team analyzed all the documentation and asked questions as necessary of the design-team members. They questioned the feasibility of the distributed-database design but after discussion, concluded that it was workable. As a precaution, however, they asked that the central database be designed to allow for the addition of where-stored information for each warehouse. This would eliminate the need for a distributed database if the proposed design proved troublesome and would limit the warehouse systems to remote printing of picking tickets and reports. While this might not be the optimum use of those systems, it at least provided a defined fallback position.

With this change, the review team certified the proposed system design. They then participated in a very brief (probably too brief) review which the design team held with the users' management. This review presented the results of the technical review and asked for management approval to continue with the project. Approval was given, and detailed system design began on the order-entry phase (refer to Figure 16-1). At the same time, the new manual methods were being implemented, so that those methods, including the use of the new order form (see Figure 16-4) would be universally in place by the time phase-in of the computer-based system began.

EXPERIENCE WITH THE SYSTEM

Although it is outside the scope of this case study to describe the system implementation, it is pertinent to discuss the company's experience with the system when operational. That experience was generally good; the system was well accepted and probably meant the difference between success and failure in handling the company's increased order volume.

System implementation of the first phase was close to schedule and was ready only 1 month later than planned. Phasing-in the systems in the warehouses, however, took longer than planned and caused more disruption in operations than expected—largely because the stock clerks' jobs changed considerably and there had been less than adequate training. However, because the distributed system was implemented at one warehouse at a time, beginning with the one located in the same city as headquarters, the disruption to the system as a whole was minimized.

The benefits expected were realized, and even though the cost of implementation was 15 percent greater than expected and ongoing costs ran 10 percent higher than estimated (largely because of a rate increase by the telephone company), the ROI was still positive. In this particular case, management would probably have accepted a negative ROI to obtain the other benefits.

Since the initial implementation, query and reporting facilities have been added. Order entry has not yet been changed from after the fact to immediate, although that is still perceived as a long-range goal. Additional uses for the minis in the warehouses are being considered; one of those is office support, to provide functions such as word processing and electronic mail to the warehouse managers' staffs.

In summary, this was a successful system analysis, design, and implementation. Although the IRM staff was under a great deal of pressure from management to implement the system as quickly as possible, they followed good practices and avoided a "quick and dirty" implementation, with the inevitable later reimplementation. Management, for its part, was very helpful in setting clear goals and in supporting IRM efforts. Every successful on-line system reflects close cooperation among the IRM staff, the users, and the appropriate members of management. That level of cooperation, combined with the use of the orderly methods described in this book, will lead to success in any complex information system.

APPENDIX

DATA-COLLECTION FORMS

The forms to be used in data collection must be appropriate to the users, the application, and the environment. The same *general* types of data are required in all cases, but the *specific* data needed will vary from application to application. The purpose of the data-collection forms is to provide a convenient way to gather and initially record the relevant data. (More permanent documentation is also required, as described in Chapter 5.) The data elements required follow:

- Output requirements

- Usage modes

- Processing requirements

- Geographical locations

- Data-storage requirements

- Integrity, security, privacy, and auditability requirements

- Expected change

Form A-1 is an example of how the basic data can be collected and documented. Although in general the system designers ought to fill out the data-collection forms, the examples shown could also be filled out by the users. Each form must be designed to be self-explanatory, as shown in Form A-1, which provides an example of the type of data needed to answer each question. The forms used in a specific organization must be customized, using examples understandable to the users.

The first form defines the functions of the user(s) and the information

DEPARTMENT: _____

FUNCTION: _____

NAME(S): _____

Please describe all the data elements you work with, by filling out this form. A data element is any item of information, such as a customer name or address, order number, and so on. Please use as many copies of this form as necessary. If the data elements you describe are included on existing business forms, please attach a copy of each form.

Data element	Size—number of letters and numbers	How it is used

Form A-2.   Detailed Data Requirements

DEPARTMENT: _____

FUNCTION: _____

NAME(S): _____

Please describe any special procedures which are used to protect certain types of data. (For example, customer names are not disclosed to outsiders to protect the customers' privacy.) Please use the same names for data elements (for example, "customer name") as on Form A-2. Please use as many copies of this form as necessary.

Data element	Method of protection/reason for protection

Form A-3. Supplemental Requirements

DEPARTMENT: _____

FUNCTION: _____

NAME (OPTIONAL): _____

Please tell us, in as much detail as you wish, what you don't like and what you do like about the computer system. If you have ideas for how to improve it, please tell us. Your suggestions can help us to improve the system.

GOOD POINTS: _____

_____

_____

_____

_____

BAD POINTS: _____

_____

_____

_____

_____

SUGGESTED IMPROVEMENTS: _____

_____

_____

_____

_____

Form A-4.   User Suggestions

GLOSSARY

Administrator, system A person who defines, controls, and manages the information-system environment within which the information-processing activities of an enterprise are handled. Specializations within the system-administrator class include database administrator and network administrator.

Advanced Communications Service (ACS) A data network planned for future operation by American Telephone and Telegraph Company (AT&T).

Advanced Research Projects Agency (ARPA) An agency of the U.S. Department of Defense, which engages in research projects; the organization which funded development of the ARPA computer network (ARPANET); now renamed the Defense Advanced Research Projects Agency, or DARPA.

After journal A record of database content following updating, to be used for recovery in case of damage to the storage medium.

American National Standards Institute (ANSI) A nongovernmental organization in the United States which defines and publishes standards related to information processing and data communications. ANSI represents the United States in activities of the International Standards Organization (ISO).

Application A definable set of tasks to be accomplished as part of the work of an organization; e.g., payroll, inventory control. An application may be partly manual and partly computerized.

Application program A set of statements defining certain tasks associated with an application that are to be executed by a computer.

Architecture A standard set of rules for functional modularity, interfaces, and protocols, which forms the framework for the implementation of products that can operate compatibly together.

Archival storage The maintenance of data for an indefinite (usually lengthy) period of time, generally with a low frequency of access. These data are seldom updated.

Area (database)   A defined subset of a database, typically used to segment the total set of data for placement on different devices, media, and/or computer systems; most often used in connection with network-structured databases.

Associative retrieval   The process of retrieving data by matching all fields in all records against a desired selection key or keys, even though the key(s) may not be used in sequencing the data for storage.

Asynchronous transmission   A mode in which the receiving mechanism synchronizes at the start and end of each character; also called start/stop transmission because a start bit precedes and one or more stop bits follow the information bits for each character.

Automated-teller machine (ATM)   A device by means of which an authorized bank customer can obtain banking services such as savings/checking withdrawal without a teller in attendance.

Auditability   The condition which allows a system to be formally investigated for correctness.

Availability   The probability that an information system, or some portion of an information system, will be usable when needed; typically computed by combining the reliability of the system components and analyzing component interdependencies.

Backup   Procedures, methods, and/or equipment which are used when the normal operating procedures, methods, and/or equipment fail.

Bandwidth   The range of frequencies assigned to a data-communications link, which determines the volume of data that can be transmitted within a given time period.

Batch processing   A mode of computer use in which input data (transactions) are accumulated into groups of convenient size before processing.

Before journal   A record of database content prior to updating, to be used for recovery in case the updating process is unsuccessful.

Binary synchronous communications (BSC)   A protocol for data communications defined by IBM, used in their products as well as in many other vendors' products; also called bisynch; being replaced by synchronous data link control (SDLC).

bps   Abbreviation for bits per second, the most common way of describing the transmission capacity of a link.

Broadband   A communications link with a bandwidth greater than a voice-grade circuit, capable of data-transmission speeds of more than 19,200 bits per second.

Call establishment   The process of connecting one endpoint through a network to another endpoint. In voice communications, call establishment is done by means of dialing or touching the desired telephone number. In data communications, dialing and touching, as well as other protocols, are also used.

Carrier   A company which, in the United States, supplies voice-, record-, or data-transmission services to the public, under control of the Federal Communications Commission (FCC) or other regulatory body. Carriers are classified as "common" or "specialized common" according to the type(s) of services offered.

Cathode-ray tube (CRT)   A term in common use to refer to a keyboard/display terminal, whether or not the display is actually a cathode-ray tube; synonymous with video-display terminal (VDT).

CCITT   See International Consultative Committee on Telephony and Telegraphy.

Centralized information system   An information system in which all processing logic is located at one site.

Circuit-switched network   A communications network in which each message is transmitted by establishing a physical connection—a circuit—between the sender and receiver and retaining that connection until transmission is complete. The voice-telephone networks are, in general, circuit-switched networks.

Common carrier   See Carrier.

Communications network   The collection of transmission facilities, modems, network processors, etc., which provides for data movement among terminals and information processors; also called data-communications network.

Concentration   The collection of data from a number of links onto a smaller number of higher-capacity links, with the distribution of the return flow from fewer to more links.

Concentrator   A network processor which performs the concentration function.

Conceptual schema   A description of a collection of entities, attributes, and relationships which represents a model of some portion of an enterprise. In general, only part of a conceptual schema is translated into a database schema, which describes the data stored in a computer-aided information system.

Customization   The process of modifying a general set of procedures and/or input methods, output methods, formats to more specifically meet the needs of individual users or user groups.

Cyclic redundancy check (CRC)   A mode of error detection used in some protocols, such as high-level data link control (HDLC).

Database   A generalized, integrated collection of company- or installation-owned data which fulfills the data requirements of all applications that access it and which is structured to model the natural data relationships which exist in an enterprise.

Database administrator (DBA) The person or group of people responsible for definition of an organization's database(s) and for the monitoring and control of operations against the database(s).

Database management system (DBMS) The functions which support the semipermanent storage of user-owned data and provide access to those data.

Database refreshing See database synchronization.

Database structure The field, record, and interrecord relationships which collectively define the format in which data are stored. The content of the structure is made up of data elements.

Database synchronization The process of ensuring that the component parts of a distributed database are logically consistent; also called database refreshing.

Datagram A technique for sending messages in a data network in which each message or packet contains all the information necessary to reach its destination and is handled independently of all other messages or packets. See also virtual circuit.

Data-sensitivity classifications The degree of importance associated with the protection of specific data elements. In the U.S. government, classifications such as confidential, secret, and top secret define how each type of data is to be protected.

DDP See distributed data processing.

Decentralized information system Two or more sets of information-processing equipment operated by the same organization but without any implied coordination among the sets.

Decryption The decoding of data which had previously been encoded, to return the data to the normal form. See also encryption.

Dedicated link A communications link which is continuously available to the user. Also called private or leased.

Dial link A communications link which is shared with other users and therefore available only when not already in use. Access is achieved by dialing or touching; also called switched.

Distributed database A single logical database which has been implemented in more than one physical segment attached to more than one information processor.

Distributed data processing (DDP) A commonly used term which is generally synonymous with distributed information system.

Distributed information system, or Distributed system A coordinated set of information-processing capabilities implemented in two or more relatively independent resource centers such as computer sites, intelligent terminal locations, and so on.

Distributed processing A technique for implementing one logically related set of information-processing (application-related) functions within multiple physical devices.

Downline load (or Download) The transmission of a program or data from one processor (usually a host) to another (usually a satellite) for execution and/or retention there.

Electronic funds transfer (EFT) The movement of funds between organizations in electronic form rather than on paper (e.g., checks), using communications facilities.

Electronic mail The transfer of textual information (messages, letters) in electronic form rather than on paper. The receiver may obtain a hard copy (a telegram is one form of electronic mail) or view the message on a display terminal.

Electronics Industry Association (EIA) A trade association in the United States which defines and publishes standards such as those required to interconnect computer and communications equipment. There is considerable overlap between the activities of EIA and of the International Consultative Committee on Telephony and Telegraphy (CCITT).

Encryption The process of converting data into an encoded form, undecipherable except to a processor or person possessing the correct decryption formula. See also decryption.

Endpoint A logical or physical entity at the end of a branch of a network. Endpoints can include application processes, terminals, and people at terminals.

End user A person who uses an information system during the performance of his or her normal duties. End users include bank tellers, retail store clerks, managers, engineers, factory workers, etc.

Ergonomic factors Aspects which affect the person/machine interface; includes items such as minimizing glare on terminal screens, determining the height and size of working surfaces for terminal users, and so on.

Factory automation The use of machines or series of machines, especially robots, to perform manufacturing tasks with little or no human intervention.

Fallback A procedure or facility which provides for continued operation, perhaps in degraded mode, when failures occur.

Fast-track A mechanism which allows an experienced user of a terminal-based system to bypass the tutorial explanations needed by inexperienced users.

Federal Communications Commission (FCC) A board appointed by the President of the United States under the Communications Act of 1934, having the power to regulate all interstate and foreign electrical communication systems originating in the United States.

File   A named increment of storage; also, an unstructured or user-structured form of data storage; often restricted to one user or one application and possibly existing only as long as that user or application is active.

Fixed-frame video   Mode in which each picture displayed on a television screen is transmitted individually and with pauses of seconds or minutes between transmissions. Each picture remains displayed until replaced by the next, so that no motion is shown.

Flow control   The function of preventing overload within a data-communications network by limiting data flow according to some algorithm(s); also called pacing.

Forms mode   A terminal-usage method in which a form is displayed on the screen so that the user can fill in the blanks on the form, resulting in the input of the required data elements.

Full-motion video   Mode in which objects on a television screen move normally. Home TV programs use full-motion video.

Function key   A specially defined key on a terminal device which, when depressed, transmits a code to the application program which processes the input. Function keys can be used to select options, define the transaction type being entered, or abbreviate input by using one key to represent a number of characters.

Global database   See system database.

Hierarchically distributed processing   A distributed system in which the logical relationships among the components—information processors, terminal controllers, real-time device controllers, etc.—form a graded series, or hierarchy; also called vertically distributed processing.

Hierarchical network   A data-communications network structure which forms a graded series, or hierarchical, pattern of connections. The simplest form of hierarchical network is a star network.

High-level data link control (HDLC)   A standard protocol for data communications, defined by the International Standards Organization (ISO).

Horizontally distributed processing   A distributed system in which the information processors cooperate in an equal partnership; also called peer-distributed processing.

Host processor   An information processor which provides supporting services and/or guidance to users and/or to satellite processors and terminals. A host processor is generally assumed to be self-sufficient and to require no supervision from other processors.

Hybrid distributed processing   A distributed system which includes both hierarchically distributed processing and horizontally distributed processing.

Institute of Electrical and Electronics Engineers (IEEE)   A professional society which, in addition to other activities, defines standards for certain computer-related subjects such as local-area networks.

Information processing   The hardware and software functions which provide computation, decision making, and data manipulation, supporting the execution of computer-aided applications.

Information resource management (IRM)   The concept of combining the responsibility for all an enterprise's systems, procedures, processing, and communication—whether manual, mechanical, or computer-aided—in one organizational entity; a broadening of the data-processing department to include functions such as voice communications, word processing, and the definition of manual methods.

Information system   An interconnected set of hardware and software components, configured to meet (some part of) the application work load requirements of an enterprise. May or may not include communications facilities.

Integrity   The prevention of, and/or recovery from, failures and errors in such a way that data and/or processes within an information system are neither lost nor damaged.

Interface   A set of rules by which services can be requested and provided. In layered architectures, an interface defines how one functional layer can request services from the next lower layer and how that layer must respond. See also protocol.

International Consultative Committee on Telephony and Telegraphy (CCITT)   An association of communications carriers and postal telephone and telegraph authorities (PTTs), which recommends common methods of transmission and interconnection.

International Standards Organization (ISO)   A multinational body which formulates standards related to information processing and data communications.

IRM   See information resource management.

Journal   A location for the storage of information retained for later recovery and/or auditing.

LAN   See local-area network.

Leased link   Synonymous with dedicated link.

Line   A link implemented as a physical connection, such as wire or group of wires, as contrasted to a microwave or communications-satellite link.

Link   The interconnection between two nodes of a network. A link may consist of a data-communications circuit or of a direct-channel connection (such as a cable or bus).

Load leveling   The function of allocating a given work load evenly over some number of processors or transmission facilities.

Local-area network (LAN)   A network which transmits data over coaxial cable, fiber optics cable, or similar media, and is limited in the distance it can traverse. A network of this type typically does not extend beyond 1.5 kilometers.

Log   A record, usually kept sequentially by time of occurrence, of events which may be of interest later. For example, a log may contain all transactions received and/or processed, the record of all actions taken by a system administrator, and so on.

Loosely coupled system   A system which includes two or more processing elements which exchange data only infrequently and generally operate independently of one another. See also tightly integrated system.

Menu mode   A terminal-usage mode in which several choices are displayed on the screen so that the user can select the appropriate option. That may result in the display of a second menu with further choices for another selection by the user.

Meshed network   A data-communications network structure in which the links provide multiple connections between any two processors.

Microcomputer, or Micro   A computer built around a "processor on a chip."

Microwave   Any electromagnetic wave in the radio frequency spectrum above 890 megacycles per second.

Minicomputer, or Mini   A "small-scale" information processor, capable of operating in office or hostile environments; i.e., with minimal environmental control.

Multipoint link   A data-communications circuit which allows connection of more than one terminal device (or possibly more than one information processor); can be configured as a loop multipoint or a multidrop.

Multiplexing   The support on a single physical link of two or more logical data streams.

National Bureau of Standards (NBS)   An agency of the U.S. Department of Commerce involved in setting standards in many areas, including certain computer-related areas.

Natural language   A terminal-usage method in which the user enters queries and/or commands in a format as close as practical to the normal spoken or written language (e.g., English, French, Russian, etc.).

Network processing   The hardware and software functions which support the definition, establishment, and use of facilities for data movement among (usually physically separated) information-system components.

Network processor A hardware device or set of devices under the control of a single set of operating-system software which provides the functions needed to control data-transmission facilities.

Network-structured database A database in which entities and relationships are defined, with related entities linked together into sets. The methods for defining and using databases of this type were described by the Conference on Data Systems Languages (CODASYL); sometimes called CODASYL databases.

Network switch A computer (usually a mini or micro) which manages a communications network, providing functions such as routing and load leveling. A network switch is a specific form of network processor.

Node An endpoint of any branch of a network, or a junction common to two or more branches of a network. In an information network, nodes includes information processors, network processors, terminal controllers, and terminal devices.

Office automation Techniques which provide computer-based assistance to people working in offices. In general, tasks are not fully automated (see factory automation) but, rather, computer-aided.

Open Systems Interconnection (OSI) The definition of a standard, layered architecture for distributed systems. The OSI standard was defined by the International Standards Organization (ISO).

Pacing See flow control.

Packet A transmission increment in a packet-switched network.

Packet assembly/disassembly (PAD) A device, usually microprocessor-based, which breaks up messages into packets (disassembly) and groups packets into messages (assembly) for a terminal connected to an X.25 network.

Packet-switched network A data-communications network in which messages to be transmitted are separated into packets (usually short and of fixed length), each of which may travel a different route to reach the desired destination.

PAD See packet assembly/disassembly.

Partitioned database A distributed database formed by splitting up the total set of data elements and attaching the resulting partitions to two or more processors.

Password A private code which must be supplied by a user or device to satisfy a predefined access-control validation procedure.

PDN See public data network.

Peer-distributed processing See horizontally distributed processing.

Personal file A collection of data which is owned by one person or by one group of people and which is not generally shared with other users.

Personal identifying number (PIN) A code known only to the user who is identified by that code and to the computer system which the user is authorized to access. Generally used in conjunction with financial systems, such as ATMs, EFT, etc.

Point-to-point A network link which connects only two endpoints to each other.

Poll and select The method by which a processor controls a multipoint link, polling to obtain input and selecting to send output.

Postal, telephone, and telegraph authority (PTT) A general term used to refer to the government authority which, outside the United States, controls voice-, record-, and data-transmission services. The authority may or may not have this exact title.

Privacy The ability to protect data pertaining to one user or user group from access, use, or modification by other users.

Protocol A set of rules for the exchange of information. In layered architectures a protocol defines rules for cooperation between two copies of the same functional layer, each resident in a different processor. See also interface.

Prototype An original model of something to be constructed; may be a model of a hardware device or system, a software program or system, and/or the user interfaces to an information system.

Public data network (PDN) A network primarily designed for data transmission and intended for sharing by many users from many organizations. Analogous to the public voice (telephone) networks.

Redundancy The configuration of additional equipment (usually hardware but sometimes software) which will be used only if the main operating equipment fails.

Relational database A database in which the relationships among entities have been reduced to the simplest possible form through a process called normalization. The relationships can be viewed as comprising two-dimensional tables, rather than sets as in network-structured databases.

Reliability The ability of an information-system component (hardware or software) to operate continuously without failure. Typically expressed in terms of mean time before failure (MTBF) and mean time to repair (MTTR). Reliability is one of the elements required to calculate availability.

Replicated database A distributed database formed by copying at least some of the data elements at two or more processors.

Requirements The functions to be provided by a computer-aided information system, as well as the frequency, availability, security, and other descriptive parameters which define the users' needs.

Resilience The ability of an information system or some part of that system to continue operation even though certain failures occur. Also called survivability.

Response The speed with which output follows input. Typically measured from the depression of the "transmit" (or equivalent) key until the first characters or line of the reply is visible at the terminal.

Resource An information-system component (hardware or software) which can serve a user requirement. Sample resources are databases, disk devices, language compilers, processes, etc.

Resource sharing The ability to make unique resources (which exist at only one or selected locations) accessible by users or applications at other locations.

Return on investment (ROI) A method for evaluating the financial payback of a new computer-aided application.

Ring network A network which has the general form of a loop, to which information processors and terminals (and sometimes other devices) can be connected. Ring networks are usually local-area networks.

ROI See return on investment.

Routing The function of selecting the appropriate path(s) for the movement of data within a network, ensuring that all data are directed to the appropriate destination(s).

Ruggedized A specially packaged computer or terminal. The packaging makes the equipment relatively insensitive to extremes of temperature, humidity, atmospheric pollution, and vibration.

Satellite, communications A device which, when placed in orbit around the earth, reflects back transmissions directed to it. All ground stations tuned to the satellite receive all transmissions. See also terrestrial.

Satellite processor An information processor which is arbitrarily assigned a subsidiary role in a distributed system, communicating with—and perhaps to some degree depending on—a host for supporting services and/or guidance.

Security The function of controlling access to, and/or the use of, resources within an information system.

Specialized-communications common carrier A common carrier which provides only certain specialized services rather than a full range of voice, data, and/or record services as provided by the conventional common carriers.

Speech recognition The ability of an electronic device to accept voice as input and decode the analog waveforms into a digital representation of the appropriate digits, letters, and words.

Standards Formally agreed-on models or rules. In the computer industry, a variety of industry and government organizations define standards. In the majority of cases compliance with the standards is voluntary.

Star network   A data-communications network structure in which links radiate from a central information processor to surrounding terminals and/or terminal controllers; the simplest form of hierarchical network.

Survivability   Synonymous with resiliance.

Switched link   Synonymous with dial link.

Synchronous transmission   A mode in which the receiving hardware synchronizes at the block level by recognition of special character(s) preceding the transmission block.

Synchronous data link control (SDLC)   A protocol for data communications defined by IBM and used in their products and in certain other vendors' products; similar to high-level data link control (HDLC).

System   A collection of components which is under a single unified management.

System administrator   See administrator, system.

System database   The total collection of data to be stored by the computerized portion of the information system; also called a global database. In the actual implementation, the system database may be segmented into several independent databases, database partitions, and/or files.

Tariff   In the United States, a definition of a specific service or services to be offered by a carrier, with the associated cost to the user. Tariffs for interstate services must be approved by the Federal Communications Commission (FCC), while intrastate or local services are approved by state or local agencies.

Teleconferencing   A method for carrying out a meeting among people who are physically in different locations. The simplest form of teleconference is a conference call. More advanced forms include not only audio facilities but high-speed facsimile, perhaps an electronic blackboard (which immediately reproduces images drawn on a blackboard in one location at other locations), and either fixed-frame or full-motion video.

Teletext   See viewdata.

Terminal   A device which embodies a set of interface functions between people and systems, such as a teleprinter or keyboard-and-display cathode-ray tube (CRT) device.

Terminal controller   A device which provides detailed control for one or more terminal devices.

Terrestrial   Land-based, as contrasted to communications-satellite-based.

Tightly integrated system   A system consisting of two or more processing elements which exchange data frequently and which operate in close coordination. See also loosely coupled system.

Time sharing   An interactive mode of computer use in which multiple terminal users concurrently access the same set of computer resources.

Training mode   A terminal-usage method which allows an inexperienced user to experiment with the system and learn its operation without interfering with other users or affecting any of the operational data in the system.

Transaction   An event (sale, order, payroll change, etc.) of interest to the organization. Also, a computer-processable record of that event.

Transaction processing   An interactive mode of computer use in which each event of interest (i.e., a transaction) is entered into the computer for processing at the time that the event occurs.

Transmission facilities   Links, switching centers, and all other equipment by means of which a communications common carrier or postal, telephone, and telegraph authority (PTT) provides a stated type of service.

Trial design   A design for a system or subsystem which must be evaluated to determine if it satisfactorily meets the system requirements. If so, the trial design can be accepted as the final design. In complex systems, several trial designs are typically formulated and evaluated before a final design is chosen.

Turnaround   The time required from the submission of input for a batch job until the resulting output is received by the submitter.

Upline dump   The transmission of memory content, following a problem or failure, from one processor (usually a satellite) to another (usually a host) for problem analysis.

User   A person known to the information system by means of a unique identifier (name, password, PIN, etc.), who can act in one or more roles (end user, application developer, system administrator) and who has a defined set of access rights to system resources.

Value-added network (VAN)   A data network operated in the United States by a specialized common carrier which obtains basic transmission facilities from the common carriers (e.g., the Bell System), adds "value" such as error detection and sharing, and resells the services to users.

VDT (Video-display terminal)   Synonymous with cathode-ray tube (CRT).

Videotex   See viewdata.

Viewdata   A method of obtaining computer-based services and/or data using a television set or TV-like terminal; also called videotex or teletext.

Virtual circuit   A technique for sending messages in a data network, in which a logical connection (the virtual circuit) is established between the sender and receiver. The physical transmission path may vary (depending on the implementation), but related messages or packets are associated with the virtual circuit until the logical connection is terminated. See also datagram.

Voice-grade A communications link which is suitable for the transmission of speech; i.e., in the frequency range of 300 to 3400 hertz; capable of data transmission at speeds up to approximately 19,200 bits per second with appropriate modems.

Wide-area network A network which can cover unlimited distances and which uses data-communications facilities for transmission; i.e., telephone lines, microwave channels, communications-satellite channels, or similar methods. See also local-area network.

Work station A multifunction terminal which usually includes multiple media, such as both a display screen and a hard-copy printer.

X.21 A definition of how terminals and computer equipment can be connected to public circuit-switched networks. Defined by the International Consultative Committee on Telephony and Telegraphy (CCITT).

X.25 A definition of how terminals and computer equipment can be connected to public packet-switched networks. Defined by the International Consultative Committee on Telephony and Telegraphy (CCITT).

BIBLIOGRAPHY

Becker, Hal B.: *Functional Analysis of Information Networks*, Wiley, New York, 1973.

Bernstein, A. J., and **A. Shoshani:** "Synchronization in a Parallel Access Data Base," *Communications of the ACM*, Vol. 12, No. 11, November 1969, pp. 604–607.

Bigelow, Robert, and **Susan Nycum:** *Your Computer and the Law*, Prentice-Hall, Englewood Cliffs, N.J., 1976.

Booth, Grayce M.: "Distributed Data Bases," *Distributed Data Processing*, Infotech State of the Art Report, Infotech International, Ltd., Maidenhead, England, 1977.

———: "Distributed Information Systems," *Proceedings of the National Computer Conference*, New York, 1976, pp. 789–794.

———: *The Distributed System Environment*, McGraw-Hill, New York, 1981.

———: *Functional Analysis of Information Processing*, Wiley, New York, 1973.

Champine, G. A.: "Six Approaches to Distributed Databases," *Datamation*, Vol. 23, No. 5, May 1977, pp. 69–72.

Chu, W. W.: "Performance of File Directory Systems for Data Bases in Star and Distributed Networks," *Proceedings of the National Computer Conference*, New York, 1976, pp. 577–587.

Connell, John J.: "Information Resource Management," *Business Week*, No. 2732, March 29, 1982, pp. 69–115.

Constantine, L., and **Edward Yourdon:** *Structured Design*, Prentice-Hall, Englewood Cliffs, N.J., 1979.

Crane, Janet: "The Changing Role of the DP Manager," *Datamation*, Vol. 28, No. 1, January 1982, pp. 97–108.

des Jardins, R., and **George White:** "ANSI Reference Model for Distributed Systems," *Proceedings of the IEEE COMPCON '78,* pp. 144–149.

Diffie, W., and **M. Hellman:** "Privacy Authentication: An Introduction to Cryptography," *Proceedings of the IEEE,* Vol. 67, No. 3, March 1979, pp. 397–427.

Erskine, S. B.: "Access to Packet Switching Networks," *IEEE Digest COMPCON '76,* IEEE Computer Society, Piscataway, N.J., Fall 1976.

Feistel, H., W. A. Notz, and **J. L. Smith:** "Some Cryptographic Techniques for Machine to Machine Communications," *Proceedings of the IEEE,* Vol. 63, No. 11, November 1975, pp. 1545–1553.

Fedida, Sam, and **Rex Malik:** *The Viewdata Revolution,* Halsted Press/Wiley, New York, 1979.

Howe, Charles L.: "Coping with Computer Criminals," *Datamation,* Vol. 28, No. 1, January 1982, pp. 118–128.

Jackson, M. A.: *Principles of Program Design,* Academic Press, 1975.

Jones, Cliff B.: *Software Development—A Rigorous Approach,* Prentice-Hall, Englewood Cliffs, N.J., 1980.

Karzan, Harry, Jr.: *Distributed Information Systems,* Petrocelli Books, New York, 1979.

Kindred, Alton, R.: *Data Systems and Management,* Prentice-Hall, Englewood Cliffs, N.J., 1973.

Martin, James: *Computer Data-Base Organization,* Prentice-Hall, Englewood Cliffs, N.J., 1975.

————: *Design of Man-Computer Dialogues,* Prentice-Hall, Englewood Cliffs, N.J., 1973.

Meadow, C. T.: *Man-Machine Communication,* Wiley, New York, 1970.

McGowan, Clement L., and **John R. Kelley:** *Top-Down Structured Programming Techniques,* Petrocelli/Charter, New York, 1975.

Palmer, Ian: *Data Base Systems: A Practical Reference,* Q.E.D. Information Services, Inc., Wellesley, Mass., 1975.

Parker, Donn B.: *Computer Security Management,* Reston Publishing Co., Reston, Va., 1981.

Parker, Donn B., and **Susan B. Nycum:** *Criminal Justice Resource Manual on Computer Crime,* SRI International, Menlo Park, Calif., 1980.

Patrick, R. L.: "Decentralizing Hardware and Dispersing Responsibility," *Datamation,* Vol. 22, No. 5, May 1976, pp. 79–84.

Scharer, Laura: "Pinpointing Requirements," *Datamation,* Vol. 27, No. 4, April 1981, pp. 139–151.

Small, D., and **W. Chu:** "A Distributed Data Base Architecture for Data Processing in a Dynamic Environment," *Proceedings of the IEEE COMPCON '79,* IEEE Computer Society, Piscataway, N.J., Spring 1979.

Tominaga, H., S. Tajima, and **K. Saito:** "Trade-off of File Directory Systems for Data Base," *Proceedings of the Fourth International Conference on Computer Communications,* Kyoto, Japan, September 1978, pp. 405–410.

Weinberg, G. M.: *Psychology of Computer Programming,* Van Nostrand Reinhold, New York, 1971.

Withington, Frederic G.: "Coping with Computer Proliferation," *Harvard Business Review,* Vol. 58, No. 3, May–June 1980, pp. 152–164.

Yourdon, Edward: *Design of On-Line Computer Systems,* Prentice-Hall, Englewood Cliffs, N.J., 1972.

———: *Techniques of Program Structure and Design,* Prentice-Hall, Englewood Cliffs, N.J., 1975.

Zelkowitz, Marvin V., Alan C. Shaw, and **John D. Gannon:** *Principles of Software Engineering and Design,* Prentice-Hall, Englewood Cliffs, N.J., 1979.

Computerworld Extra! Special Edition: "Data Communications: Making Critical Connections," Vol. XVI, No. 11a, March 17, 1982.

Guidelines for ADP Security and Risk Management, U.S. Department of Commerce, National Bureau of Standards, FIPS Publication 31, U.S. Government Printing Office, Washington, D.C., 1974.

INDEX

ABOUT THE AUTHOR

Grayce M. Booth is the Manager of Communications for Honeywell Information Systems in Phoenix, Arizona. She has been in data processing since the late 1950s, with a career spanning an unusually wide range of areas in programming, system design, system engineering, and marketing. She has worked on application system development, operating system design and implementation, database management, and networking system software design.

Since 1974 Ms. Booth has been involved with distributed systems, working with users and prospective users of Honeywell large-scale systems and minicomputers, defining market requirements, product marketing, and assisting in the product planning of distributed system capabilities. During the last two years she has included office automation functions in the scope of her distributed system activities.

Ms. Booth is the author of numerous technical articles dealing with distributed systems, networking, database management, and other topics. She has lectured in the United States, Europe, Japan, and Australia. She is the author of two books: *Functional Analysis of Information Processing*, published in 1973 by Wiley-Interscience, and *The Distributed System Environment*, published in 1981 by McGraw-Hill.